Fodé Yansané
Daouda Konaté

Metal pollution

Fodé Yansané
Daouda Konaté

Metal pollution

Studies of metallic pollution on some species of fish caught in Tabounsou-Conakry bay Tabounsou-Conakry

ScienciaScripts

Imprint
Any brand names and product names mentioned in this book are subject to trademark, brand or patent protection and are trademarks or registered trademarks of their respective holders. The use of brand names, product names, common names, trade names, product descriptions etc. even without a particular marking in this work is in no way to be construed to mean that such names may be regarded as unrestricted in respect of trademark and brand protection legislation and could thus be used by anyone.

Cover image: www.ingimage.com

This book is a translation from the original published under ISBN 978-620-6-72612-8.

Publisher:
Sciencia Scripts
is a trademark of
Dodo Books Indian Ocean Ltd. and OmniScriptum S.R.L publishing group

120 High Road, East Finchley, London, N2 9ED, United Kingdom
Str. Armeneasca 28/1, office 1, Chisinau MD-2012, Republic of Moldova, Europe
Printed at: see last page
ISBN: 978-3-330-07716-4

Dedication

With enormous pleasure, an open heart and immense joy, I dedicate this memoir to my dear,
respectful and magnificent parents, the late Ibrahima Sory and Mamaïssata **TOURE**, as a sign of
love,
recognition and gratitude for all the sacrifices they have made on my behalf.
May your soul rest in peace, dear Papa.

I also dedicate this work to my big sister M'Mahawa **YANSANE** and my big brothers Mikaël,
Alhassane and Ibrahima Sory **YANSANE**, who came to my aid during the most
difficult moments
of my training. Thank you so much; God will keep you by our side for a long time.

♦ To everyone who helped me and to my friends at ISSMV/Dalaba.

♦ And to all those who are dear to me.

TABLE OF CONTENTS

FOREWORD

Like many cities around the world, Conakry is currently experiencing problems with waste management (domestic and industrial) and maritime waste (from ships and aircraft).

For local residents, the coastline is regarded as a rubbish dump. Wastewater from all sources is discharged into the sea without prior treatment. The liquid effluents discharged by most of the thermal power stations, as well as rainwater, are evacuated by a single evacuation system made up of several secondary collectors that meet up in a canal, which drains them into outlets that open directly into the sea, causing pollution.

Pollution from land-based sources, maritime transport and dumping at sea, as well as a large number of polluting substances, pose particular problems for the marine environment in that they combine toxicity, persistence and bio-accumulation in the food chain. In this complex environment, fish are able to concentrate useful elements (iron, copper and calcium) during their feeding, but also toxic elements (lead, mercury, arsenic) that can be responsible for digestive, nervous and skin disorders.

Metals such as Pb, Ni, Hg, Zn, Cd and Cu from industrial wastewater can harm the quality and survival of aquatic species.

It was with this in mind that, at the end of our university course, we chose the topic ***"Studies of metal pollution in some species of fish caught in the Tabounsou-Conakry bay"***. During the course of this work, we encountered difficulties linked to the innovative nature of the topic, our young experience in scientific research, and the lack of material and financial resources, which prevented us from carrying out this work to the height of our ambitions. Nevertheless, we remain hopeful that, after criticism, suggestions and amendments, our results will serve as a guide for future researchers interested in this field.

We owe the credit for this work to God the Almighty for giving us the life, health and courage to carry it out. This work is the fruit of a collective effort, and we would like to express our sincere thanks to our consultants Dr Daouda **KONATE** (PhD), Dr Abdoulaye **BAH** (PhD) and Mr Mohamed Boly **CAMARA**, respectively Head of the Department of Veterinary Medicine at Dalaba, Secretary General of the Doctoral School of the Conakry Rogbanè Scientific Research Centre (CERESCOR), Head of the Maritime Fisheries Chair at Dalaba and Mr Ousmane II CAMARA, Fisheries Biologist at the Centre de Recherche Scientifique Conakry Rogbanè (CERESCOR). Ousmane II **CAMARA**, Fisheries Biologist at the Centre National des Sciences Halieutiques de Boussoura (CNSHB), for their rigour, guidance, attention and wise advice in carrying out this work.

We would like to express our deepest gratitude to all those who have contributed in any way to the preparation of this memoir.

The candidate

SUMMARY

This descriptive research on the theme of ***"Studies of metal pollution on some species of fish caught in the bay of Tabounsou-Conakry"*** was carried out from 25 September 2020 to 14 March 2021 inclusive.

The aim of this study was to assess the level of contamination of some fish species by trace metal elements (TMEs) in Tabounsou Bay with a view to proposing mitigation measures for these elements.

To achieve this objective, we adopted the following methodology:

1. Consultation of frameworks and analysis of archives ;

2. On-site survey ;

3. Sampling ;

4. Determination of the physico-chemical parameters of the water ;

5. Determination of metal content in water and in the flesh of fish species ;

6. Proposed mitigation measures.

The consultation of managers, analysis of archives and the field survey revealed not only the causes of the pollution (diversification of waste, dumping of hydrocarbons at sea) and its impact on the ichthyofauna (alteration of fish) in Tabounsou Bay through galloping demographics, but also the list of fish species landed at the various sites *(Pomadasys jubelini, Aruis latiscutatus, Ethmalosafimbriata, Mugil cephalus, Dentex angolensis, etc.)...).*

Analysis of the physico-chemical parameters of the water at the sampling sites (Faban, Gbessia port 2 and Gbessia port 1) gave the following results respectively: Temperature 33.82; 22.4; 23.9°C; Salinity 33.82; 23.9; 32%; pH 6.2; 6.8; 6.5; Oxygen 5.2; 6.8; 5.8 mg/l; Turbidity 32.3; 16.7; 31.8 NTU; Conductivity 73; 61; 69 mS/cm.

Assessment of the metal content of the water at the sampling sites (Faban, Gbessia port 2 and Gbessia port 1) gave the following results:

- ➤ For Zn (1.58; 1.38; 1.45) and Cu (1.28; 1.12; 0.82) mg/l ;
- ➤ For Pb (1.02; 0.3; 0.46) and Cd (0.31; 0.024; 0.027) mg/l.
- ➤ Evaluation of the average metal content in the fillets of the three species of fish also gave respectively (expressed in mg/l) :
- ➤ For *Pomadasys jubelini* Zn (70 and 61.3); Cu (22 and 19); Pb (0.4 and 0.34) and Cd (0.04 and 0.016);
- ➤ For *Mugil cephalus* Zn (30 and 26); Cu (32.5 and 31); Pb (0.6 and 0.49) and Cd (0.07 and 0.055);
- ➤ For *Dentex angolensis* Zn (24.5 and 22.8); Cu (29.6 and 28); Pb (0.21 and 19) and Cd (0.067 and 0.03).

Demographic growth and the development of human activities along Tabounsou Bay are having a detrimental impact on fish stocks.

INTRODUCTION

Preserving the marine and coastal environment is becoming an ever-growing concern, as more than half of humanity lives close to the sea and draws most of its resources from it.

However, these resources, which are beneficial to human health, are being jeopardised by human activities (industrial effluents, domestic effluents and run-off, inputs of metals from rural areas, etc.), oil spills following shipping accidents and effluents from thermal power stations, including used oil, grease and fuel oil sludge, which also pose a serious threat to coastal and marine ecosystems. Added to these anthropogenic activities are the metals emitted as a result of natural inputs: volcanoes and forest fires (Rocher V., 2003).

In Africa, wastewater is generally untreated and discharged directly or indirectly on or near the seafront, often in estuaries, or even in the immediate vicinity of towns. As a result, local populations are involuntarily exposed to nuisances and cardiovascular infections such as bronchitis, asthma and gastritis (Wognin SB., 2008).

In Guinea, urban infrastructures are far from ensuring basic conditions of cleanliness, due to the lack of urban sanitation and waste treatment facilities, which are leading to a deterioration in the quality of coastal ecosystems. Today, it is recognised that the increase in built-up areas in the coastal zone, particularly in Conakry and the secondary towns, clearly indicates that the prospects for demographic growth are strong. Sources of hydrocarbon and heavy metal pollution in Guinea come from industrial zones, seaports, shipping lanes and thermal power stations (Bah A., 2019).

Thus, the proliferation of sources of discharge through industrial effluents, septic tanks, household waste and run-off water channels in the southern part of Conakry (Tabounsou Bay) is likely to promote the dysfunction of the coastal ecosystem, with the immediate consequence of the modification and migration of biological diversity and the reduction of biological resources and risks of the appearance of epidemiological and contagious diseases (such as cholera) on a large scale (Bah A., 2019).

The pressures exerted on this environment are leading to a reduction in its biodiversity. A good understanding of this ecosystem is therefore essential. It is for these reasons that we have decided to focus our scientific research on the theme of ***"Studies of metallic pollution on some species of fish caught in the Tabounsou-Conakry bay"***.

To carry out this study, we have structured our dissertation report as follows:

CHAPTER I: LITERATURE REVIEW ;

CHAPTER II: MATERIALS AND METHODS ;

CHAPTER III: RESULTS AND DISCUSSION.

CHAPTER I: LITERATURE REVIEW

I. General information on marine pollution

European legislation defines pollution as "the direct or indirect introduction, as a result of human or natural activity, of substances or heat into the air, water or soil which may adversely affect human health or the quality of aquatic ecosystems or terrestrial ecosystems directly depending on aquatic ecosystems, which result in damage to material assets, deterioration of or interference with amenities or other legitimate uses of the environment" (Ouali N., 2018).

Pollution is any anthropogenic and natural modification of an ecosystem resulting in a change in the concentration of natural chemical constituents, or resulting from the introduction into the biosphere of artificial chemical substances, a disruption in the flow of energy, the intensity of radiation, the circulation of matter or the introduction of exotic species into a natural biocenosis (Wassim G. 2017).

Pollution has also been defined as: "the introduction by man into the environment of substances or energy likely to cause damage to the health of living species (humans, living resources and ecological systems)". According to the same author, contamination is defined as: "an anthropogenic input of pollutants into the environment, but without any harmful effects on the health of living species" (Sahbaoui F., 2015).

1.1. Definition of marine pollution

Marine pollution is the alteration of the aquatic environment: a change in the state of an aquatic environment or a water system, leading to degradation. Alterations are defined by their nature (physical, ionic, organic, toxic and bacteriological, etc.) and their effect (bloom, asphyxiation, poisoning, modification of populations, etc.) (Malquiot M and Bertolini P H., 2000).

Pollution can have repercussions at all trophic levels, from primary producers to higher consumers, and can therefore affect the functioning of ecosystems (Triquet- Amiard C., 2008).

The term pollutant has been defined as a biological, physical or chemical alteragen (element), which above a certain threshold or standard, develops negative impacts on all or part of an ecosystem or the environment in general. (Gouery D, 2014).

1.2. Types of marine pollution

a) Biological pollution

This pollution is divided into two types:

■ Bacterial pollution

This is the most important problem from an economic and health point of view, and concerns beach pollution (water and sand), skin diseases and gastroenteritis in bathers (Cravez V and Bernard G., 2006).

■ Pollution by marine species foreign to the marine environment

Pollution can be caused by the introduction of a marine species into an area where it is normally absent (invasive species) and where it has a significant impact (Bouchriti P., 2003).

b) Physical pollution

Physical pollution occurs when the physical structure of the marine environment is altered by various factors, such as the release of freshwater to reduce salinity (by a hydroelectric power station), the release of heated or cooled water (by a power station or factory), the release of liquid or solid substances that alter the turbidity of the environment (mud, silt, macro-waste, etc.), or the release of radioactivity (Cravez V and Bernard G., 2006).

c) Chemical pollution

Chemical pollution of water results from the release of certain toxic mineral substances into watercourses, such as nitrates, phosphates, ammonia and other salts, as well as metal ions (Pb, Ni, Hg, Zn, Cd, Cu). These substances have a toxic effect on organic matter, making it more dangerous. The result is radioactive pollution, where the radioactivity of natural waters may be of natural or artificial origin (nuclear energy) (Cravez V and Bernard G., 2006).

d) Eutrophication of the coastline

Nutrients will contribute to the proliferation of aquatic plants and algae. Although the latter form the basis of the food web in aquatic ecosystems, their proliferation goes against the natural balance between producers and consumers, and gives rise to other problems such as bloom and asphyxia (Berg D M. et *al.,* 2009).

I.3. Origins of marine pollution

The production and emission of pollutants are often derived from human activities, such as agriculture (fertilisers, pesticides and agrochemicals), industry (trace elements and organic compounds), urban development (pathogens, organic substances, trace elements in waste water) and tourism (litter, plastics on the coast) (Larno V., 2001).

a) Urban effluents

Most urban effluent discharges directly into the immediate marine environment without any treatment measures. This effluent is heavily laden with mineral pollutants and micro-organisms (bacteria, pathogenic viruses and parasites). Within urban agglomerations, these effluents contain chemical waste from both household and industrial activities, and in the majority of cases flow towards main collectors. Terrigenous inputs during precipitation can also drain (household) waste into aquatic environments through leaching (Sahbaoui F., 2015).

b) Industrial effluents and emissions

Industries located in coastal areas discharge their waste directly into the sea or into watercourses. Atmospheric emissions from industry project pollutants that can be transferred by air to the sea (Sahbaoui F., 2015).

Industrial waste can contain heavy metals (copper, cadmium, nickel, mercury, lead, etc.) or other hazardous substances (hydrocarbons, titanium, etc. ....) (Chiffoleau J F et a/.,2001).

c) Rivers and streams

Polluted rivers carry considerable loads of liquid waste into the marine environment. In addition to inputs from industrial facilities and urban agglomerations, rivers can sometimes also carry fertilisers and pesticides used in agriculture. Rivers thus make a major contribution to the transport of pollutants (Sahbaoui F., 2015).

d) Uncontrolled coastal discharges

Several forms of solid and liquid waste deposition in or near the marine environment contribute directly or indirectly to sea pollution, depending on the type and quantity of material deposited (Sahbaoui F., 2015).

e) Maritime transport

Oil spills from shipping accidents, including used oil and grease, also pose a serious threat to coastal and marine ecosystems. Maritime transport is best known to the general public for the spectacular damage and pollution it inflicts on the sea. It is estimated that 4 to 6 million tonnes of oil arrive or are spilled into the oceans every year, and the French Institute for the Environment reports that 600,000 tonnes of hydrocarbons are spilled every year in the Mediterranean, and three million tonnes every year in the North Sea (Chiffoleau J F et *a/.*, 2001).

I.4. Causes of marine and coastal pollution

a) Spills from thermal power stations

The liquid effluents discharged by most thermal power plants, as well as rainwater and fuel oil sludge containing metals (Hg, Zn, Cu, Pb, Ni, etc.), are evacuated by a single drainage system that drains them into outlets that open directly into the sea, causing pollution (Wassim G., 2017).

Petroleum products (oil and gas) accounted for 65% of total energy consumption in the United States in 2000, thus covering most of the country's energy needs, which alone consume *25%* of the commercial energy used in the world as a whole. The total mass of fossil energy burnt in 2000 is equivalent to 11% of the total amount of solar energy fixed each year by photosynthesis by all the primary producers present in continental ecosystems. The main air pollutants produced by combustion are S02 and NOX, which result in acid rain, photochemical smog and solid particles (Wassim G., 2017).

b) Diversification of chemical pollutants and accumulation of waste

Since the end of the Second World War, organic chemistry has provided us with a multitude of new synthetic molecules. In the 1990s, it was estimated that 120,000 molecules were marketed worldwide, and that between 500 and 1,000 new chemical substances were brought onto the market every year. These include plastics, detergents and insulating materials, all of which have undeniably made everyday life more comfortable. One of the most formidable aspects of global pollution by synthetic chemicals is the increase in discharges of persistent organic pollutants (POPs) into the marine environment in both industrialised and developing countries. Because of their ubiquity and stability, these substances (PAHs, PCs, organochlorine pesticides, dioxins, etc.) are now found in the most remote regions of the biosphere (Wassim G., 2017).

c) Pollution from agriculture and intensive livestock farming

A final major cause of pollution of the biosphere is the development of a model of agriculture and animal husbandry that is reputed to be modern. The use of synthetic pesticides (insecticide, fungicide, herbicide) and mineral fertilisers (nitrogen, phosphate and potassium salts) in intensive agriculture has led to spectacular progress in crop yields. These products, in addition to livestock excrement, generate insidious pollution of surface and coastal waters and groundwater (Wassim G., 2017).

I.5. Sources of pollution in the Guinean coastal zone and marine waters

The main sources of pollution of the coastal zone and marine waters are diverse and varied. There are four main categories of sources of pollution from land-based activities that affect Guinea's coastal and marine environment (Camara T and Diallo ST., 2006).

a) Urban and industrial wastewater

They are one of the main factors contributing to the degradation of the marine and coastal environment. It is well known that, apart from the separate sewage network in the commune of Kaloum and a few installations in the town of Kamsar (Boké), most sewage is discharged via septic tanks and landfill in the city of Conakry, where 80 to 90% of the country's industrial units are located. The Conakry slaughterhouse discharges between 3,000 and 7,000 tonnes of solid and liquid waste into the sea every year. What's more, all kinds of products that pollute marine waters (rubbish and/or various domestic wastes, detergents, hydrocarbons, chemical residues from industries, clinical and hospital wastes, slaughterhouse wastes, etc.) find their way into the sewers. This leads to significant changes in water quality (pH, conductivity, hardness, dissolved oxygen, nitrite/nitrates, phosphates, etc.) and has a dangerous impact on marine fauna and flora. As well as posing a health risk to consumers, this type of pollution also leads to eutrophication of the water, leading to blooming, the development of pathogenic bacteria and changes to the biotic and abiotic parameters of the environments in question. Conakry currently has a wastewater decantation project covering an area of 25 hectares. In the town of Fria, there is a system for decanting wastewater (Doté) from alumina processing before it is discharged into the Konkouré river. In Guinea's coastal regions, there are no wastewater treatment or pre-treatment facilities (Camara T and Diallo ST., 2006).

With regard to mining activities in the coastal zone, the following should be noted:

- The Compagnie des Bauxites de Guinée (CBG - Kamsar) extracts and processes 11,000 to 13,000 tonnes of bauxite per year. The dust from the crushing and pre-treatment process covers the whole of the locality's southern greenery, which is washed into the coastal waters as soon as the first rains fall. The crushing of bauxite and the handling of hydrocarbons at the port give rise to muddy water and other effluents that end up in the sea. Most of the used oil is discharged into the marine environment. Other environmental problems include dust flying around Kamsar and the production of solid and liquid waste (water consumption is estimated at 9,000 m^3 /day) in the towns of Kamsar and Sangarédi. An estimated 6,300 m^3 of wastewater is discharged every day.

- The Friguia Kimbo alumina plant, which is an industrial unit, extracts bauxite ore and processes it on site into alumina, which is transported to the Port Autonome de Conakry (PAC). The effluent from the processing, which used to be discharged into the Konkouré river, could be seen as far away as Conakry. Nowadays, the effluent is discharged into a settling tank near the factory. During the alumina production process, products such as sulphur, lime and soda are used at various stages. The quantity of red mud resulting from the separation of sand suspensions is estimated at 700 - 800 tonnes per year. The sludge is drained into the Doté lake for settling before being discharged into the Konkouré river. According to some studies, this plant discharges one tonne of sludge per tonne of alumina produced, and each tonne of red sludge contains around 15 kg of soda (Na OH), which is not reabsorbed by the process. In addition, this red mud is composed of 60% iron or (Fe2 O3), 30% carbonates ($CaCO_3$) with traces of Titanium (Ti O_2) (Camara T et al., 2006).

a) **Agrochemical waste**

These are forms of pollution of agricultural origin, including pesticides, herbicides and persistent organic pollutants. They generally predominate around irrigated areas (Camara T and Diallo ST., 2006).

According to the results of a survey carried out by the National Customs Department, Guinea imports almost 11,000 tonnes of mineral fertilisers. These imports rose from 6,925 tonnes in 1995 to 10,990 tonnes in 1998. Most of the imports come from Côte d'Ivoire (74.2%), followed by France (11.3%) and Japan (8.9%). Triples 15 and 17, urea, ammonium sulphate, super triple and potassium sulphate are the types of fertiliser most frequently imported. Excess fertilisers infiltrated into surface or groundwater also contribute to the enrichment of water and, consequently, to its eutrophication potential, especially during the rainy season (Camara T and Diallo ST., 2006).

b) **Rubbish, solid waste, plastics and marine debris**

Rubbish, solid waste, plastics and marine debris are generally found in large cities such as Conakry, Dubreka, Boké and Coyah, all of which are located near the coast or adjacent to river catchment areas. The city of Conakry had a population of around 2,160,000 in 2005, and the tonnage of waste generated per day is of the order of : 750-800 t/d, a daily production per inhabitant of : 45-50 kg/day, i.e. 155 to 160 kg/year/inhabitant. The La Minière landfill receives an annual tonnage of around 200,000 t, with a collection rate for the landfill of 70 to 75%. Rubbish is distributed in three ways along the coast:

- direct dumping of rubbish on the coast: the quantity is small (less than 10%) and its position is homogeneous or not very varied, some is biodegradable and others is not (kitchen leftovers) on the seashore in coastal towns (Camara T and Diallo ST., 2006).

- redistribution of rubbish from the coast by the tide: the quantity of rubbish in this category is significant (over 30%). Its composition is very varied and the materials are non-biodegradable (plastics, fabrics, old tyres, old metal sheets, etc.).

- drainage of rubbish by run-off water: plant debris and other abandoned materials are carried away by run-off water as it passes. It should be noted that some coastal residents take advantage of heavy rainfall to use urban road drains as rubbish dumps. The various types of waste carried by run-off make up more than 60% of the solid waste found in the coastal zone. Analyses carried out by the National Environmental Laboratory and the Ministry of Health Laboratory show that the water along the Guinean coast has coliform concentrations of 2.4 to 80 times the WHO standard. The services currently provided to treat this waste consist of collecting it and transporting it to uncontrolled landfill sites, which are generally located in the middle of urban areas and have no development plan. The mining company's landfill site in Conakry, which covers an area of 2 hectares, is surrounded by housing. The environmental and safety problems posed by this landfill are enormous (contamination of surface water bodies, loss of historical and recreational value, degradation of marine and coastal ecosystems) (Camara T and Diallo ST., 2006).

c) **Mining**

Guinea is a country rich in a variety of mineral resources, which are exploited by industrialists and craftsmen alike. It is estimated that Guinea's bauxite reserves, which have not yet been fully explored, exceed 10 billion tonnes, making the country the richest in bauxite in the world.

Bauxite deposits are mainly found in three geographical areas: Boké-Gaoual and Kindia-Fria, in Lower Guinea, and Dabola-Tougué in Upper Guinea. Reserves in the Boké-Gaoual area are the largest, accounting for around two-thirds of national reserves.

In the Boké area, bauxite occurs on easily accessible plateaux, in layers around 8 m thick. It is almost entirely free of waste rock. The coastal ecosystem is home to four (4) mining ports and numerous mining sites. Industrial mining activities exert heavy pressure on the fauna, flora, freshwater and soil in these Prefectures. It disturbs the soil, destroys vegetation, degrades landscapes and destroys low-lying areas (farmland par excellence), and dumps red mud into watercourses. This sludge fills in estuaries, rivers and ponds, causing pollution and problems with the availability of drinking water due to the permanent turbidity of the water bodies, as well as serious problems for the survival of biological diversity and populations. Today, the landscapes of Lower Guinea are very much marked by the vast bleeding of open-cast mines and by the considerable destruction of the soil, ecosystems, plant cover and fauna. Many social groups have benefited from the conversion of natural ecosystems as a result of the exploitation of mining resources. However, these benefits have been realised at increasingly high costs, in the form of loss of biodiversity, degradation of many ecosystem services and increased poverty for other social groups (Bah .M 2015).

Although local people have benefited from mining activities, the costs of the changes borne by the countryside are often higher. In many cases, it is particularly important to note the loss of income that mining generates for local populations due to the destruction of farmland and livestock meadows, the destruction and/or pollution of watering holes, the destruction of habitats and dwellings, not to mention the evictions, dislocation of families, socio-cultural losses, etc. (Bah.M 2015).

Table I.1: Nature and quantity of pollutants discharged into the sea in Guinea.

Nature of pollutants	Quantity (tonnes per year)
Oil	600
Lubricants	867
Solution of inorganic bases	110,000 at sea and on land
Non-biodegradable basic plastics	599
Combustible organic matter from :	
-From the industry	105
-Craftsmanship	1905
- Households (household waste)	8600
- Wrecked vehicles (scrap metal)	1200

Source: National Environmental Action Plan, 1994

I.6. Consequences of pollution

The consequences of pollution can be classified into three main categories (Khelil F., 2007).

a) Health consequences

The impact of pollution depends on the concentration of pollutants, their virulence, the duration of exposure and the amount of physical effort involved. These four factors are very important in accurately assessing the health risks associated with pollution in an individual.People who bathe in water polluted by sewage often suffer from gastrointestinal disorders, ear infections, eye and skin infections and respiratory problems. Cholera epidemics and viral hepatitis are common among coastal populations, and each time they occur they result in a large number of fatal cases (Khelil F., 2007).

b) Aesthetic consequences

They disrupt the image of an environment (for example, plastic bottles or tar washed up on a beach) (Khelil F., 2007).

c) Economic consequences

Economic losses for commercial fisheries in certain regions where fishing and marine culture have had to be limited or abandoned for public health reasons, or fish stocks have been reduced as a result of the destruction of habitats or spawning grounds. The decline in the quality and quantity of fish products from developing countries (Khelil F., 2007).

Table I.2: Diseases linked to marine pollution and economic impact worldwide

Diseases linked to contamination of the marine environment	Economic impact/year (billions of dollars)
Related to bathing and swimming	1,6
Seafood consumption (hepatitis)	7,2
Seafood consumption (algae toxins)	4,0
Subtotal	**12,8**

Source: State of the environment and policies from 1972 to 2002

d) Social and cultural consequences

At a social level, the city of Conakry is conducive to tourism because of the beaches that border it (Bel Air), the estuaries (Tabounsou estuary, Soumbouya estuary, Konkouré estuary etc.) that cut into it and the islands (Loos Islands, Alcatraz Island, etc.) (Direction Nationale du tourisme et hôtellerie Conakry, 2014). Indeed, the islands of Kassa, Soro, and Loos Island receive many changing rooms for maritime discovery. Unfortunately, this traffic causes its share of problems, as the more people there are, the greater the chance of disturbance. High visitor numbers mean that the nightlife in towns and cities is more active, with tourists frequenting restaurants and bars, increasing noise levels and disrupting the local population. Indeed, boats and pirogues can often obstruct the view and uglify the landscape, as they are massive and disrupt the natural aspect of the landscape (Direction Nationale du tourisme et hôtellerie Conakry, 2014).

I.7. Effects of pollution on fish and wildlife

Pollution of aquatic environments has many consequences. Physico-chemical alterations involve changes to the characteristics of the environment, such as salinity, pH or water temperature. Beyond a certain threshold, these changes become toxic for the organisms living in the environment (Http//www.eaufrance.fr/les impacts de la pollution de l'eau, 2020).

Of all the physico-chemical parameters, oxygen and temperature are particularly decisive for fauna and flora. A quantity of dissolved oxygen that is too low to sustain life is known as hypoxia. Anoxia is the final stage, when there is no more dissolved oxygen in the water. Episodes of hypoxia can be the result of too much organic matter. This is broken down by the bacteria in the environment, which consume the oxygen dissolved in the water during this process. However, hypoxia can also be caused by other factors, such as an increase in water temperature (oxygen being less soluble in water) (Http//www.eaufrance.fr/les impacts de la pollution de l'eau, 2020).

This is particularly true of heavy hydrocarbons (crude oil and other types of fuel oil), which have multiple impacts linked to the physical engulfment they cause: the vegetation covered up is suffocated, oiled birds are unable to fly and can no longer feed, while light hydrocarbons such as fuels have more toxic effects. Waste can also harm biodiversity, particularly fish fauna, which confuse waste with food. Plastic waste can also have toxic effects or cause endocrine disruption (Https://www.eaufrance.fr/les impacts of water pollution, 2020).

Nutrients (phosphates, nitrates) are generally present in limited quantities in aquatic environments and constitute what are known as limiting elements. Any additional supply of these elements is rapidly assimilated and stimulates primary production. When the natural cycle is disrupted by human activities, in particular by fertilisers, detergents and waste water in general, excess phosphates (and to a lesser extent nitrates) are responsible for the phenomenon of eutrophication. This results in an excessive proliferation of algae and/or macrophytes, and a reduction in water transparency. The decomposition of this abundant organic matter consumes a great deal of oxygen and more often than not leads to the mass death of animal species through asphyxiation (Leveque C et *al,* 2006).

The ecological consequences of the proliferation of invasive plant species are no less significant: they lead to the asphyxiation of aquatic environments through the process of eutrophication. In addition, these aquatic plants compete with native plant species, which are excluded from their natural habitat (Hill K., 2003).

I.8. Waste

The definition of waste is set out in the French law of 1975, which initiated France's waste management policy. Waste is defined as "any residue of a production, transformation or use process, any substance, material, product or, more generally, any movable asset that is abandoned or that its holder intends to abandon".This definition must be supplemented by the more restrictive European Directive of 18 March 1991, which considers waste to be "any substance or object included in the European Waste Catalogue which the holder discards or intends or is required to discard". This definition reflects the historical and social nature of waste through the idea of "abandonment", which marks a reduction in value, a downgrading, a relegation to the margins (Christian N and Alain R., 2004).

a) Types of waste

Human activities produce large quantities of waste: 886 million tonnes in France in 2006, with a wide variety of characteristics. Depending on its origin, waste can be grouped into four major families containing a wide variety of residues (Christian N and Alain R., 2004).

-Municipal waste: this includes household refuse, residual sludge from wastewater treatment plants, waste from green spaces and waste from government departments, businesses and local authorities; cleaning product residues, paint residues, expired medicines, animal waste (mainly used to spread on land), animal carcasses, packaging (plastic, cardboard, paper).

Waste from agriculture and the agri-food industry: this includes pesticide and fertiliser residues, oils, parts of agricultural machinery and equipment, and construction debris.

-Industrial waste: which includes waste oils, sludge from reservoirs or industrial processes, ash and bottom ash from industrial boilers.

Waste from healthcare activities: which includes human anatomical waste (including blood), whether infectious or not, animal anatomical waste (laboratories), non-anatomical and infectious waste: syringes, dressings, residues (Christian N and Alain R., 2004).

From an environmental point of view, waste is a threat; from an economic point of view, it is a potential source of wealth. Not all waste poses a threat to human health; the main problem is how it is managed. Only toxic, infectious or radioactive waste requires stringent management precautions (Hill K., 2003).

The new European directive of 19 November 2008, which will repeal the 1975 directive on waste oils and the 1991 directive on hazardous waste on 12 December 2010, also specifies that "any waste presenting one or more of the hazardous properties listed in the directive (irritant: inflammatory reactions on the skin and mucous membranes; harmful: limited risks through ingestion, inhalation or skin penetration; toxic: serious or even chronic risks through ingestion, inhalation or skin penetration) must be considered hazardous".Hazardous waste includes specific waste from businesses, dispersed toxic waste (DTQD) produced by all kinds of structures, including healthcare establishments, and hazardous household waste (DDM) (Hill K., 2003).

II. Metal pollution

Trace metal pollution represents a real problem that depends on the environment, the physiological state of the organism and certain environmental factors (temperature, pH...) (Mansouri K et *al.,* 2015).

Living organisms are selective about the metal load in their bodies. The elements Na, K, Mg and Ca are present in large quantities because they have an essential role in metabolic functions (major elements) whereas other metals are present in much lower concentrations (trace elements); Among the trace elements, there are those that are essential to life (Cu, Zn, Co, Mn, Fe, Al, Mo, Si, V) and those that are not (Cd, Pb, Hg, Sb, As, Ba, Be, Se, Ag) (Mansouri K et *al.,* 2015).

II.1. Definitions of heavy metals

There are many different definitions of heavy metals, depending on the context and the purpose of the study. From a purely scientific and technical point of view, heavy metals can also be defined as :

Any metal with a density greater than 5,

Any metal with a high atomic number, generally higher than that of Sodium (Z=11),

Any metal that can be toxic to biological systems.

Some researchers use even more specific definitions. Geologists, for example, consider any metal that reacts with pyrimidine to be a heavy metal (Ouali N., 2018).

In liquid waste treatment, the undesirable heavy metals we are mainly interested in are: As, Cd, Cr, Hg, Ni, Pb, Se, Zn (Ouali N., 2018).

In the environmental sciences, the heavy metals associated with the notions of pollution and toxicity are generally: As, Cd, Cr, Cu, Hg, Mn, Ni, Pb, Sn, Fe, Zn (Ouali N., 2018).

Finally, in industry in general, a heavy metal is considered to be any metal with a density greater than 5, a high atomic number and posing a danger to the environment and/or humans (Ouali N., 2018).

11.2. Biological classification of heavy metals

The classification of heavy metals is often debated, as some toxic metals are not particularly 'heavy', such as zinc. On the other hand, some elements are not metals but metalloids, such as arsenic.

For all these reasons, most scientists prefer the term "trace metals" (TMEs) to "heavy metals", or by extension "trace elements" (Miquel, 2001).

a) Essential trace elements

Essential metals are trace elements that are essential for many cellular processes and are found in very low proportions in biological tissues. These trace elements must meet the criteria laid down by (Cotzias, 1967).

- be present in living tissue at a relatively constant concentration;

- cause structural and physiological abnormalities due to their absence in the body

- prevent or cure disorders by providing this threshold element.

b) Non-essential trace elements

Non-essential metals have no known beneficial effect on the cell, but are pollutants with toxic effects on living organisms, even in low concentrations, such as lead (Pb), mercury (Hg) and cadmium (Cd) (Chiffoleau, 2004).

11.3. Sources of trace metal contamination

Trace metals are natural constituents of the earth's crust, present mainly in the form of ores. They are therefore naturally present in the biosphere and in the atmosphere. Volcanism, fires and thermal springs all contribute to the release of metals into the environment (Rocher F., 2003).

These natural inputs have been supplemented by metals emitted as a result of human activities: mining of deposits and use of metals in many sectors (metallurgy, foundries, incineration of waste, combustion of fossil fuels and materials, spreading of plant protection products and fertilisers in agriculture). The transfer of these compounds is essentially due to the leaching of soil and run-off water, which drains large quantities of products and residues into aquatic areas. The contaminants can then adsorb onto mineral particles (loss of bioavailability, trapping in sediments). They are also capable of circulating in water for a more or less long time (substances that are more or less persistent) (Ouali N., 2018).

11.4. Penetration mechanisms of heavy metals in marine organisms

In marine organisms, these toxic elements penetrate via three routes (Sahbaoui F., 2015)

- **The trans-tegumentary route:** direct contamination from the outside environment.

- **The respiratory route (gills):** this is the main route of contamination.

- **Trophic pathway**: depends on the diet.

11.5. Transfers of contaminants in the marine environment

Many marine organisms accumulate contaminants in very high concentrations in their tissues. These accumulation processes depend on the assimilation, excretion and storage rates of each element (Sahbaoui F., 2015).

- **Bioaccumulation:** is a physiological mechanism that results in the fixation of toxic substances in marine organisms. It is therefore the ability of a given species to concentrate a given toxic substance from the external environment; these non-biodegradable substances will concentrate along the various links in the trophic chain, with maximum concentrations found in large predators (fish, marine mammals, humans) or in filter-feeding molluscs such as mussels.

- **Bioconcentration:** bioconcentration is a special case of bioaccumulation. It is defined as the process by which a substance (or element) is present in a living organism at a higher concentration than in its surrounding environment. It is therefore the direct increase in concentration of a contaminant when it passes from water to an aquatic organism. The concentration factor CF is defined as a constant resulting from the ratio of the concentration of an element in an organism in a state of equilibrium to its concentration in the biotope.

- **Biomagnification:** is the concentration of a toxic substance after the smallest organism in the chain has been consumed by the largest; in this case, it is the possibility of a toxic substance being accumulated by a trophic chain. If the toxic substance is not degraded or eliminated, it will accumulate more and more at each link in the food chain.

11.6. Biogeochemical cycle of heavy metals in the marine environment

There appear to be two main stages in a biogeochemical cycle.

Firstly, metal pollutants would be trapped by particles in suspension, by marine biomass and by sediment, depending on the physico-chemical conditions of the marine environment:

- **Precipitation:** phenomenon that occurs when the metallic pollutant in solution falls to the bottom of the marine environment by gravity. However, in deep water, some metals may return to solution well before reaching the bottom.

- **Absorption:** phenomenon that occurs when molecules or metal ions attach themselves to the surface of marine components (particles, marine organisms, sediments).

- **Adsorption:** this is the passage of the metallic pollutant into a marine organism.

- **Sedimentation: a** phenomenon that occurs when metal ions are superimposed, forming sediment layers. Benthic animals help to accelerate the deposition of particles and their associated metals by consolidating them in faecal matter. These animals thus contribute to the sedimentation of the marine environment.

A second stage, the reverse of the first, would consist of the release of these pollutants by desorption or sorption, which is the opposite phenomenon to adsorption, by diffusion or propagation in the marine environment by redissolution or resolution of products precipitated by decomposition and remineralisation of organic matter, and sometimes even by redistribution via marine organisms (Sahbaoui F., 2015).

11.7. Impact of heavy metals on human health

Among the mineral chemical elements, metals occupy a predominant place in our modern world because they are involved in most sectors of activity. Many of them are also essential to the living world (iron, zinc, etc.), sometimes in very small quantities (essential trace elements). Some of these trace elements (lead, manganese, etc.), which are essential in small doses, become toxic in high concentrations. Finally, there are metals such as mercury, cadmium and chromium, which are uniquely toxic to living organisms (Picot A., 2002).

Cadmium poisoning in pregnant women has been linked to a reduction in the duration of pregnancy and in the weight of the newborn, and recently to endocrine and/or immune system dysfunction in the child. Exposure to lead has repeatedly been linked to delayed neurobehavioural development. Several studies have been carried out on fertility by testing the vitality of sperm; exposure to these metals reduces this capacity. While treatment of eggs with each metal (Cd, Hg, Pb, Ni and Zn) did not prevent fertilisation, it did delay or block the first mitotic divisions, and an early change in embryonic development is envisaged (Lidsky T I and Schneider J S., 2003).

II. 8. Toxicity of heavy metals on fish

There are three types of toxicity, depending on the speed of onset, severity and duration of symptoms and the speed of absorption of the toxic substance (Sahbaoui F., 2015).

- **Acute toxicity:** This occurs during a very short period in the life of an organism, through rapid absorption of the toxicant, by the transmembrane route (if the species is injured), the gill route (breathing) or the buccal route (feeding). Manifestations of intoxication develop rapidly, and death, very serious physiological disorders or recovery occur without delay.

- **Sublethal toxicity:** In this case, frequent or repeated exposure over a period of several days or weeks is required before symptoms appear.

- **Chronic toxicity:** This is manifested by the toxic effects produced not by the absorption of fairly large doses over a short period of time, but rather by exposure to very low concentrations, sometimes even minute doses, of polluting substances in the repetition of cumulative effects that are measured on the parameters that are generally more sensitive, such as reproduction and behavioural changes (physiology). This chronic toxicity ends up causing much more serious disorders.

III. Studies of some heavy metals

III.1. Zinc

Zinc is a bluish metallic chemical element, symbol Zn and NA 30. It is a so-called essential metal. Zinc is a trace element necessary for the metabolism of living beings, essential for many metalloenzymes and transcription factors that are involved in various cellular processes such as gene expression, signal transduction, transcription and replication; it also participates in the photosynthesis phenomenon for plants. It enters the atmosphere naturally from wind transport of soil particles, volcanic eruptions, forest fires and marine aerosol emissions. Man-made inputs of zinc into the environment come from industrial mining sources (ore processing, refining, iron galvanisation, roof gutters, electric batteries, pigments, plastics, rubber), agricultural spreading (animal feed, slurry) and urban activities (road traffic, waste incineration). In port areas, zinc is introduced from the dissolution of anodes used to protect ship hulls against corrosion; it is also contained in certain antifouling paints. Zinc is one of the most abundant trace elements in humans (needs 15 mg/day). It is involved in growth, bone and brain development, reproduction, foetal development, taste and smell, immune functions and wound healing (Gagneux-Moreaux S., 2006).

Long-term exposure to zinc can cause anaemia, gastrointestinal problems and diarrhoea. Zinc's toxicity to aquatic organisms does not make it a priority contaminant, although in high concentrations it affects oyster reproduction and larval growth. At high concentrations, zinc becomes toxic to plants and animals and is a major contaminant of the terrestrial and aquatic environment (Lafabrie C., 2007).

The **FAO/WHO (1989)** directive on the edibility of fish and fishery products sets the thresholds at 40 to 100 mg/kg.

111.2. Copper

A reddish-brown metallic chemical element with the symbol Cu and NA 29, it is a so-called essential metal.

The main sources of copper in coastal ecosystems are discharges of urban and industrial wastewater (metallurgy, chemicals) and leaching from agricultural soils. This trace element is used in the composition of many plant protection products, and is therefore found in the marine environment, causing disruption to species (Nakhlé B., 2005).

Copper is bio-available to organisms in oxidation state (I) or (II), from inorganic salts or organic complexes. It is involved in many metabolic pathways, including haemoglobin formation and neutrophil maturation. It is also a specific cofactor in many enzymes and structural metalloproteins involved in oxidative metabolism, cellular respiration and pigmentation. Copper is of vital importance in the maintenance of biological processes. In molluscs, the blood contains a copper-based respiratory pigment, haemocyanin. Toxic to animals and micro-organisms in doses of less than 1 mg/l, decreases photosynthetic activity (marine plants), causes damage to gills and delays egg-laying in fish; copper is responsible for Wilson's disease in humans, which is due to the accumulation of copper in the liver (Nakhlé B., 2005).

Recommendations according to the directive on the edibility of fish and fishery products set the thresholds at less than 30 mg/kg (**FAO/WHO 1989).**

111.3. Lead

One of the non-essential TMEs (Trace Metal Elements). It is a bluish-grey metallic element with the symbol Pb and atomic number 82. In the environment, lead is mainly present in the atmosphere and comes from smelters, the metallurgy industry, coal combustion, waste incineration and vehicle exhaust fumes. In the marine environment, it arrives mainly via atmospheric inputs and leaching from urbanised areas. Effects on the respiration, growth, reproduction and behaviour of vertebrates and invertebrates have been observed at much higher concentrations of the order of mg per litre. Lead in inorganic form is said to exert its toxicity by competing with metals essential to the normal functioning of the cell. Lead is currently one of the most important pollutants because of its non-degradability and its cumulative effect in natural environments and organs (Verloo P., 2003).

Exposure to low doses of lead can have certain effects on the intellectual development and behaviour of children. Exposure to high levels of lead can cause kidney disease, mental retardation, anaemia and reproductive problems. According to the same author, chronic exposure to lead can have negative cardiovascular effects on humans, and it is also a carcinogen. Anaemia is a characteristic sign of lead poisoning; children are more sensitive than adults, and the nervous system is also affected. Lead poisoning varies with the duration and intensity of exposure. In fish, lead, like copper, increases with age, accumulating in the liver, kidneys and spinal column (Casas S., 2005).

Lethal doses of lead, in the form of mineral salts, are often higher than its solubility limit in seawater, i.e. 4 mg/l. Inorganic lead can therefore be considered toxic (lethal concentration of 1 to 10 mg/l) or moderately toxic for seagrass beds, particularly eelgrass and Posidonia (lethal concentration of 10 to 100 mg/l). The regulatory health quality threshold is 1.5 mg/kg-1 under European regulation EC 221/2002.

III.4. Cadmium

It is one of the so-called non-essential trace metal elements, with the symbol Cd, NA 48 and a brilliant white colour. Cadmium is highly resistant to corrosion and has chemical characteristics similar to those of calcium, in particular its ionic radius, making it easier to penetrate organisms (Turkmen A et *al.,* 2005).

Cadmium is an element found in aquatic environments in various physical (dissolved, colloidal, particulate) and chemical (mineral or organic) forms. A range of physicochemical environmental variables (salinity, pH, redox potential, sedimentological characteristics, geochemical nature of particles, chloride concentration) govern the transformation of cadmium in the environment. Cadmium is naturally present in trace amounts in the surface rocks of the earth's crust, making it a rarer element than mercury and zinc. There are two main sources of cadmium:

- Primary cadmium is mainly associated with zinc in zinc ores (0.01 to 0.05%) and is therefore a by-product of zinc metallurgy, which yields an average of 3 kg of cadmium per tonne of zinc.

- Cadmium is also present in lead and copper ores and in natural phosphates (Jordan, Tunisia) (Chiffoleau J.F et *al.,* 2001).

Cadmium is a pollutant linked to a number of modern industrial processes, and is also produced in the agricultural region as a contaminant in phosphorous fertilisers and in sewage sludge, which is also used for fertilisation. Part of the cadmium input to coastal environments comes from the atmospheric compartment (smoke and dust from smelters, products of the incineration of cadmium-coated materials) and part from leaching from agricultural land containing fertilisers, which releases and transports cadmium and other trace elements into the marine environment. Unlike many metals, cadmium has no known metabolic role for humans and does not appear to be biologically essential or beneficial to the metabolism of living organisms. It sometimes replaces Zn in Zn-deficient enzyme systems in plankton. Regulation (EC) No 466/2001 sets maximum cadmium levels in foodstuffs (1 mg/kg wet weight). However, it is not acutely toxic to marine organisms at concentrations likely to be found in the environment. At a sublethal level, concentrations of 0.05 to 1.2 pg/l can cause physiological effects (anomalies in embryonic and larval development in bivalve molluscs) and growth inhibition (Chiffoleau J F et *al,* 2001).

IV. Studies on a number of fish species

IV.   1. *Pomadasys jubelini*

a) Definition and morphology

The Pomadasys genus is a semi-pelagic species of the Haemulidae family, with a laterally compressed body. The protractile mid-mouth is supported by a snout. The operculum has no spines. The caudal fin is forked. On each side of the fish, a lateral line continues from the operculum to halfway along the caudal fin. *Pomadasys jubelini* can be distinguished from other fish of the same genus by the height of its body, which is one third of the fork length (distance from the end of the mouth to the fork of the *caudal* fin).*Pomadasys jubelini* has 5 rows of scales above the lateral line at the origin of the dorsal fin; the last dorsal spine is longer than the penultimate; the second anal spine is stronger and longer than the third; the maxilla is massive and robust; and the spots are relatively small, sometimes more or less regularly arranged and dark brown in colour (Paugy et *al.*, 2003).

b) Habitat and Food

Pomadasys jubelini is a Haemulidae that lives in marine environments up to 100 m deep and can penetrate brackish waters and sometimes even freshwater. Its body has a silvery background with dark-brown spots scattered fairly irregularly along the rows of scales on the back and flanks; the species is considered to be an omnivore. It is one of the species of marine origin perfectly adapted to lagoon conditions (Paugy et *al.*, 2003).

c) Reproduction

Their reproductive cycle involves a series of physiological and behavioural processes linked to various factors in the biotic and abiotic environment. The reproductive effort involves not only significant energy expenditure sustained by the direct supply of nutrients, but also the use of reserves previously built up and stored in the hepatopancreas. Mastering the biology of Pomadasys jubelini therefore requires knowledge of its cycle and its energy storage organ, as the hepatopancreas is necessary for gametogenesis. Differentiation of the hepatopancreas and gonads occurs during the sexual maturity of Pomadasys jubelini (Paugy et al., 2003).

Male Pomadasys jubelini have hepatopancreas consisting of two distinct lobes. These hepatopancreas evolve through five stages of maturity. Classification is essentially based on the size, vascularisation and coloration of the lobes. These different stages are as follows:

Stage 1: Individuals measure an average of 19.6 cm with an average height of 5.2 cm and weigh 106.2g. Their hepatopancreas measures 3.3 cm with a weight of 1 g and consists of two unequal brown lobes. The left lobe is larger than the right.

Stage 2: specimens measuring 20.9 cm by 5.4 cm. They weigh 150g and have hepatopancreas measuring an average of 4.2 cm. These hepatopancreas weigh 1.56 g and are red in colour with two unequal lobes, as the right lobe is less developed.

Stage 3: Fish at this stage weigh 221.6 g. They are 23.7 cm long and 7 cm high. Their hepatopancreas are 4.6 cm long and weigh 2.45 g. The hepatopancreatic organs are bilobed and light brown in colour. The two lobes are of equal size, but the left lobe is larger than the right.

Stage 4: specimens reach 28.4 cm in length and 6.9 cm in height. They weigh 318.8 g (Figure 1D) and have garnet-coloured, bilobed hepatopancreas measuring 5.5 cm. These organs weigh 3.1 g. The lobes are unequal, with the left lobe slightly larger than the right. **Stage 5:** The individuals are 28.5 cm tall and 6 cm high. They weigh 240 g. Their hepatopancreas are 5.5 cm long and weigh 2.9 g. They are light brown in colour and consist of two asymmetrical lobes.

Table I.3: Taxonomy of the species *Pomadasys jubelini* (Ried K., 2004).

CLASSIFICATION	
Kingdom	Animalia
Branch	Chordata
Class	Actinopterygii
Order	Perciform
Family	Haemulidae
Type	*Pomadasys*
Species	*Jubelini*
Scientific name :	*Pomadasys jubelini*

French **name** : Grondeur sompat.

Name in Sousou: Kèssi-kèssi

Figure I.1: Image of *Pomadasys jubelini*

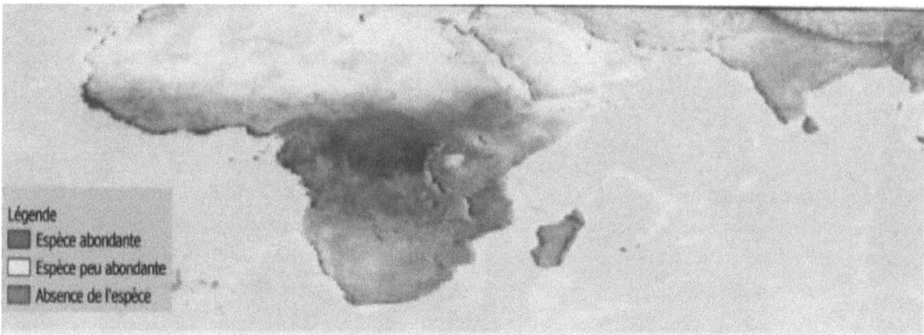

Figure I.2: Geographical distribution of *Pomadasys jubelini* in Africa

IV.   2. *Mugil cephalus*

a) Definition and morphology

The Mugil genus are coastal fish from tropical and temperate seas. The body is fusiform, cylindrical or compressed, particularly towards the rear. The head is broad and flattened, the snout short and obtuse. The mouth is small and terminal, with or without teeth; the eye may or may not be covered by a strong adipose eyelid. The scales are fairly large, but there is no lateral line. The dorsal fins are clearly separated from each other, the first consisting of 4 spines. The pectorals are high on the flanks; the pelvises are in the abdominal position and have a scaly process at their base. The caudal fin is forked or truncated. The body of the *Mugil cephalus* is robust, cylindrical in shape and slightly compressed laterally. The head is broad and depressed. The upper lip is thin and smooth, without papillae. A thick adipose eyelid almost completely covers the eye, leaving only a vertical elliptical slit over the pupil. The second dorsal fin and the anal fin are only covered with scales in their anterior part and at their base. The anal fin has only 8 soft rays. The back is bluish grey, while the flanks and belly are silvery with, in general, grey longitudinal lines and golden highlights. The anal fin and the lower lobe of the caudal fin are yellowish.

b) Habitat and Food

Mugil cephalus species form "troops" of varying sizes, moving around in search of food. Muges often move back and forth between lagoons and the sea. Migrations should be cyclical and predictable, covering more than 100 km, with females generally migrating to marine waters at sea to give birth, but returning to freshwater to feed. Muges are thalassotoic (catadromous) migratory fish, i.e. they migrate to the sea to reproduce (Ried K., 2004).

(Hill K., 2004) defines 3 types of migration: a migration of juveniles from the sea to estuaries, a breeding migration and a migration of adults to the open sea after a few years. During the autumn and winter months, adult mullet migrate far out to sea to spawn in large accumulations. They are particularly fond of brackish waters with wide variations in salinity and are abundant in estuarine and lagoon areas. Some species migrate up the lower reaches of rivers and are perfectly suited to living in freshwater, but reproduction always takes place at sea.

They are limivores, which means that they swallow the mud and sift it through their well-developed gill apparatus to extract the organic particles. They also graze on algae and small organisms on rocky bottoms around quays and harbour structures, where moving shoals can be observed. Adult mullet feed on plants and small animals (invertebrates) and suck up sediment from the seabed. Mullets often form shoals that graze on small plants attached to seaweed. Mullet are preyed upon by large fish such as the lutjan or the barracuda.

c) Reproduction

Mugil cephalus is a gonochoric species, with no visible sexual dimorphism. Breeding generally takes place from July to October; they reproduce at sea; the breeding period varies with water temperature. Sexual maturity is reached in males at three years of age, when they reach a size of between 33 and 38 cm (Collins and Stender 1989) and in females at four years of age, when they are between 40 and 42 cm. Females can lay between 500,000 and 3,700,000 pale yellow, slightly buoyant eggs with a diameter of between 0.72 and 0.78 mm. The larvae (2.4 mm) hatch 48 hours after fertilisation and drift in the plankton towards the coasts and estuaries. The general reproductive pattern of the Mugil cephalus involves migration to the sea between July and October, with females spawning between 0.5 and 2.0 million eggs per female, depending on the size of the individual. Spawning occurs in deep water from mid-October to the end of January, with peak spawning occurring in November and December. The larvae and pre-juveniles then migrate to coastal estuaries where they inhabit shallow, warm water in the inner tidal zone. *Mugil cephalus* is the most fecund of the mullet species. The larvae and pre-juveniles that migrate towards the estuary seem to respond to a combination of biotic and abiotic factors; this fish does not seem to be bothered by polluted, oxygen-poor and low-salinity waters (Bester C., 2004).

Table I.4: The average length (cm) as a function of sex inMugil *cephalus* (Ouali N., 2018).

Age	Male	Female
2 years	21.1	35.5
3 years	42.5	50.1
4 years	49.3	58.9
5 years	54.0	64.5

Table I.5: Taxonomy of the species Mugil *cephalus* (Bester C., 2004).

CLASSIFICATION

Kingdom	Animalia
Branch	Chordata
Class	Actinopterygii
Order	Mugiliformes
Family	Mugilidae
Type	Mugil
Species	Cephalus
Scientific name	*Mugil cephalus*

Figure I.3:Image of *Mugil cephalus*

Name : French :Mulet

Name in Sousou: Sèki

Figure I.4: Geographical distribution of *Mugil cephalus* worldwide

IV. 3 *Dentex angolensis*

a) Definition and morphology

Dentex angolensis is a species of the family Sparidae, a benthic pelagic with demersal behaviour. Sparidae are perciformes with a generally high and compressed body, often with a typical high frontal profile. They have a single dorsal fin with 10 to 13 spines and 10 to 15 soft rays, an anal fin with 3 spines and 8 to 12 rays and a forked caudal fin. But the essential characteristic of the Sparidae is their dental differentiation or heterodonty. In this family, the teeth are specialised according to the species' diet: herbivores such as sars, bogue and saupe have flat, sharp incisors; predators, such as dentex, have hooked canines and crustacean and shellfish eaters, such as paddlefish, have crushing molars; finally, pageots, which eat debris, have teeth similar to those of paddlefish, but less powerful. All these different types of teeth, as well as their arrangement, are used in the classification of genera. Another feature of the Sparidae is their frequent hermaphroditism: individuals can be first male and then female, like the sars (protandry) or, conversely, female and then male, like the paddlefish.

b) Habitat and Food

Dentex angolensis prefers hard, sandy bottoms, from 20 metres to 250 metres. Individuals move mainly in shoals in the first 100 metres. The distribution of the species along the coast is not uniform and muddy areas are devoid of this species. It is considered to be a predominantly carnivorous omnivore, feeding on benthic invertebrates and small fish. It lives on hard, sandy bottoms in water with a salinity of over 34%. The minimum and maximum sizes observed are 4 to 30 cm respectively. Young individuals are the most coastal: they appear in the middle of the dry season and are abundant until the start of the rainy season. They then gradually disappear.

c) Reproduction

Dentex angolensis reproduction is intermittent from the second year onwards. Mature individuals move closer to the coast when they lay their eggs. They are hermaphroditic (the majority of individuals are initially female and then become male). Mature at 2-4 years (16-35cm) with a fecundity of 60200 to 406800 oocytes (Bauchot ML et Hureau JC.,1990).

Table I.6: Taxonomy of the species Dentex angolensis (Bauchot ML et Hureau JC.,1990).

CLASSIFICATION	
Kingdom	Animal
Branch	Vertebrate
Class	Actinopterygii
Order	Perciform
Family	Sparidae
Type	Dentex
Species	Angolensis
Scientific name :	*Dentex angolensis*

French name : Denté angolais
Name in Sousou: Sinapa

Figure I.5: Image of *Dentex angolensis*

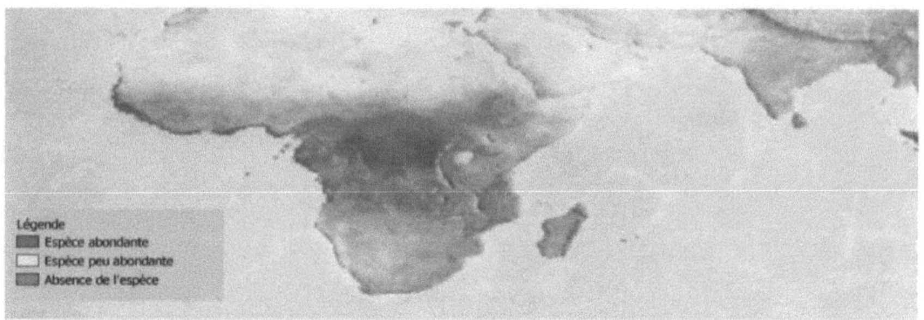

Figure I.6: Geographical distribution of *Pomadasys jubelini* in Africa

CHAPTER II: MATERIALS AND METHOD

A. Hardware

11.1. Monographic study of Conakry

1.1.1. Geographical location and population

The city of Conakry, capital of the Republic of Guinea, is located between 9° 32'53" North Latitude and 13° 40'14" West Longitude. It covers an area of 450 km^2 and stretches for approximately 34 km. It widens towards the north-east and south-east, by 1 to 6 km (Bah A., 2019).

With an altitude of up to 130 metres, the western part of the city of Conakry is made up of a flat, low-lying area extended in an easterly direction by a ridge with two slopes (south and north), describing uniform gradients of 4 to 5% on average, either directly to the sea or over flat marshy areas.

It also includes the archipelago of Loos islands (Kassa, Rome and Fotoba). The population density in this city is 3706 inhabitants/km^2 (Bah A.,2019).

It is limited :

- To the east by the prefecture of Coyah ;

- To the west by the Atlantic Ocean ;

- To the north by the prefecture of Dubréka ;

- To the south by the prefecture of Forécariah (RGPH, 2014).

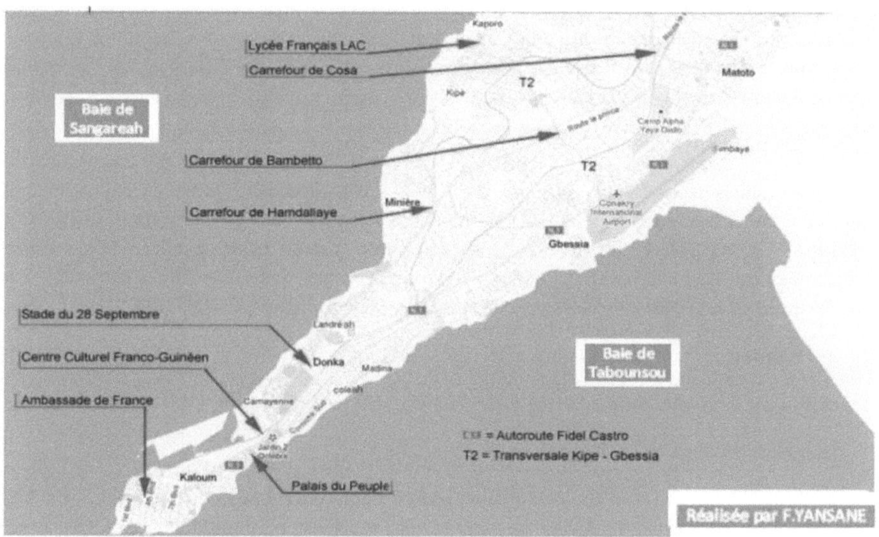

Figure II.1: Map of the city of Conakry

II.1.3. Physical data

a) Hydrography

The hydrography is characterised by the presence of a number of rivers that drain surface water towards the sea. Their course is almost straight and perpendicular to the longitudinal axis of the city. These rivers carry nutrients and heat to the sea. In addition to rainfall, this freshwater input causes desalination of the river mouths. In addition, their discharges may reduce salinity, which will cause the migration of certain species of fish, particularly stenohaline fish (Direction de l'environnement eaux et forêts-Conakry, 2014).

b) Relief

The relief of the city of Conakry is formed by a plain that runs alongside the plateaux. It takes the form of a platform that is uneven in places, giving it a slightly undulating appearance. The altitude of Mount Kakoulima (1007m) plays an important role in the phenomenon of condensation (Direction de l'environnement eaux et forêts-Conakry, 2014).

Its undulating appearance encourages the drainage of run-off water towards the sea. This drainage has both positive and negative effects. Run-off water carries nutrients to the sea and also promotes high turbidity in estuaries, a phenomenon that contributes to slowing down the development of fish (Conakry Governorate, 2014).

c) Climate

Generally speaking, the city of Conakry has a humid tropical climate known as sub-Guinean, characterised by the alternation of two (2) distinct seasons of equal duration (six 6 months); a dry season from December to May characterised by a rise in the surface temperature of the sea water as a result of sunshine and a rainy season from June to November when there is heavy rainfall which drains household waste into the sea thus causing pollution which constitutes a real threat to aquatic organisms in general in particular fishery resources (Direction Nationale de la station météorologique Conakry, 2014).

- **Temperature**

The temperatures recorded are relatively high with small differences (between day and night) varying from one season to another and this rise in temperature would have an impact on dissolved oxygen so its increase could have negative effects on aquatic fauna and flora. This rise in temperature can lead to a reduction in dissolved oxygen, which will affect fish in particular, who will be asphyxiated (Direction Nationale de la Station Météorologique-Conakry, 2014).

- **Rainfall**

Rainfall is high, ranging from 2,900 to 5,700 mm per year. The average is 4300 mm per year.

The rains last between 100 and 115 days.

Rainfall is the main driver of the desalination phenomenon (during precipitation) and of the high turbidity observed in estuaries during the rainy season, which can be a limiting factor for the life of certain marine organisms that cannot tolerate low salinities and favours that of organisms that prefer low salinities, in particular the larvae and juveniles of many organisms: fish, crustaceans, algae, etc. (Direction Nationale de la Station Météorologique-Conakry, 2014).

- **Winds**

Two types of wind are encountered in Conakry:

-The monsoon: is a cool, wet wind that blows full force against the Fouta-Djalon escarpments facing the sea, causing heavy rainfall at the foot of Mount Kakoulima (1107 m). It blows from the ocean towards the mainland, from south-west to north-east. This heavy rainfall drains domestic and agricultural waste containing various chemical substances into the sea, causing pollution that will affect the life of aquatic species and their habitat.

-Harmattan: is a hot, dry wind that almost never reaches the coast, as it is blocked by the mountain ranges of Fouta-Djalon and carried aloft by wet winds. It blows from the continent towards the ocean from north-east to south-west (Direction Nationale de la Station Météorologique-Conakry, 2014).

At sea, wind can be both necessary and disastrous for fishermen. It creates waves that can topple small boats, but it is also a means of propulsion for sailing boats. It is the main driving force behind drift currents (surface currents), which cause pollutants to disperse in the marine environment, particularly hydrocarbons, which prevent light from penetrating the environment, affecting both aquatic plants and fish stocks.

II.1.2. Socio-economic activities

- **Agriculture**

Generally speaking, farming activities are hampered by the lack of agricultural land due to urbanisation. The fertilisers and pesticides used are washed into the sea, which can cause eutrophication in these environments, and the deposition of excess plant matter at the bottom of the aquatic environment leads to bacterial growth, which consumes dissolved oxygen and causes an oxygen deficit in the environment, leading to the death of aerobic organisms such as fish (Conakry Governorate, 2014).

- **Breeding**

Livestock farming is a minor economic activity in the urban area. The most widespread type of livestock farming is poultry, which is essentially family-based. It is aimed exclusively at self-consumption. However, with the removal of barriers to liberal initiative, operators are now promoting small-scale poultry farms in the peri-urban areas of Ratoma and Matoto. The promotion of this activity on a large scale is coming up against demographic pressure. Other types of farming, such as small ruminants (sheep and goats) and pigs on a small scale, are often practised near aquatic environments (particularly near estuaries) and most of their faecal matter is discharged into these environments or drained by soil leaching during the rainy season (Section Communale de l'Elevage et Administration 2014).

- **Fishing**

Because of its proximity to the Atlantic Ocean, Conakry has a large number of landing stages: Port de Faban, Boulbinet, Kaporo, Landréya, Temenètaye, Bonfi, Dixinn Port III; and has an Autonomous Port where industrial fishing products and other goods are unloaded. Fishing is considered to be one of the main activities of the local population. It is both small-scale, semi-industrial and industrial. It is experiencing a small boom with the establishment of small fishing units: SAFRI-PECHE, Adams Pêche, Nimba Pêche, Nicolas pêche etc. Traditional offshore demersal fishing uses modern equipment with 12-14m motorised boats and 15-40hp engines. The fishing grounds are remote but always within sight of the coast, and the majority of the owners are Guinean, with a few foreigners. Today's fishing industry, with its continuous development and the diversification of the gear and boats used, is causing enormous damage to the marine environment and the resources that live there. Plastic and other waste dumped at sea by most fishermen also causes enormous damage to fish stocks, particularly pollution. In the Tabounsou estuary, an increasing decline in fish catches has been noted, with juveniles predominating, adult males and females at the reproductive stage being rare due to the use of prohibited fishing methods and nets, and regular dumping of domestic and industrial waste along the coast (Konaté, S. et *al.,* 2007).

- **Maritime transport and ports**

Maritime transport is highly developed in Guinea. Nevertheless, the operation of ships and maritime activities in general cause pollution. Apart from spills from oil tankers, other pollutant spills can be caused by shipping: used oils and lubricants from ships, spills of waste produced on board, discharges of spoiled cargo, etc., not to mention spills resulting from shipping accidents.

There are two large-capacity ports on the coast: the Port Autonome de Conakry, which handles all types of vessel (container ships, ore carriers, oil tankers, etc.), and the Port Industriel de Kamsar, which only handles ore carriers (bauxite).

As far as hydrocarbons are concerned, there have been clear cases of pollution in the port area of Conakry and to the north of Conakry. In Conakry, more than 20 major waste and sewage sites have been identified, posing a real threat both to fish stocks and to the health of the population (Plan National d'Action pour l'Environnement, 1994).

- **Floors**

Quite permeable, with a water table, Conakry's soil is varied and closely linked to the geographical structure; it is made up of lateritic soils, most of which are ferralitic.

Alongside these types of soil are alluvial soils formed from sands resulting from surface erosion in adjacent regions. These are rich soils but can become lateritic as a result of uncontrolled deforestation and exploitation.

The leaching of these non-clay soils by heavy rainfall can be a source of pollution in coastal and estuarine areas, as it causes severe turbidity that can affect the life of aquatic organisms living in the water column. In fish, suspended particles can cover the gills and prevent gas exchange with the outside environment.

In filter feeders, certain particles that may be toxic to the body can be absorbed during the absorption process (Direction de l'environnement eaux et forêts Conakry, 2014).

- **Crafts**

Handicrafts is a traditional field that brings benefits to the country's economy, unfortunately in the city of Conakry this art activity is quite interesting due to the fact that economically this sector brings less to the population of the city despite it embodies the culture of the country, this sector contains the following trades: shoemaking, dyeing, blacksmithing, sculpture and artisanal fishing, all of which produce waste that is dumped directly or indirectly into the sea, causing damage to the aquatic ecosystem and its fishery resources (Centre culturel Franco-Guinéen, 2010).

- **Industry**

The city of Conakry has a large number of functional industries in the country. Most of these industries are of the agri-food type, with the exception of a few that manufacture building materials. Some of these industries are in the process of slowing down due to a lack of investment or subsidies. Among those in operation are Bonagui, Métal Guinée, Sobragui, Topaz and ciment de Guinée.

Most industries are located by the sea. Industrial waste is always drained into the sea, and this can harm fish stocks, especially larvae, which are fragile at this stage of their development. This type of pollution poses a health risk to consumers and leads to significant changes in water quality (physical and chemical parameters), which ultimately affects marine ichthyofauna (Conakry Governorate, 2014).

- **Tourism**

Conakry has a lot to offer in terms of tourism. There are tourist sites such as Kassa Island, Soro and Los Island, but this potential is hampered by a lack of promotion, poor management and the degradation of these natural qualities by domestic waste and human activities that contaminate bathers. Hotels are one of the main producers of household waste, which is discharged into the sea and contributes to the pollution of the marine coastal environment (Direction Nationale du tourisme et hôtellerie Conakry, 2014).

II.2. Description of Tabounsou Bay

II.2.1. Geography and population

The Tabounsou estuary is located to the south-east of Conakry in the urban district of Matoto. It is bounded to the north by the urban district of Coyah, to the north-west by Conakry, to the south by the Atlantic Ocean and to the south-east by the Soumbouya river basin. The investigation area is located between 9° 24' - 9° 40' N and 13° 35' - 13° 45' W to the south of the city of Conakry.

It is supplied with freshwater by the rivers Tabounsou (main river), Tombolia, Kountia and Trente Six (km36) on the left, and the Sarinka on the right. The waters of the Toguiron, Kitéma and Soumbouya streams, located to the right towards the east only feed the bay during the rainy season (Direction Nationale de la Station Météorologique-Conakry, 2014).

The Tabounsou bay is made up of the riverside populations of the Matoto and Coyah urban districts, the majority of whom are Sousous. These two communes have an estimated population of 1003676 (Recensement général de la population et de l'habitation, 2014).

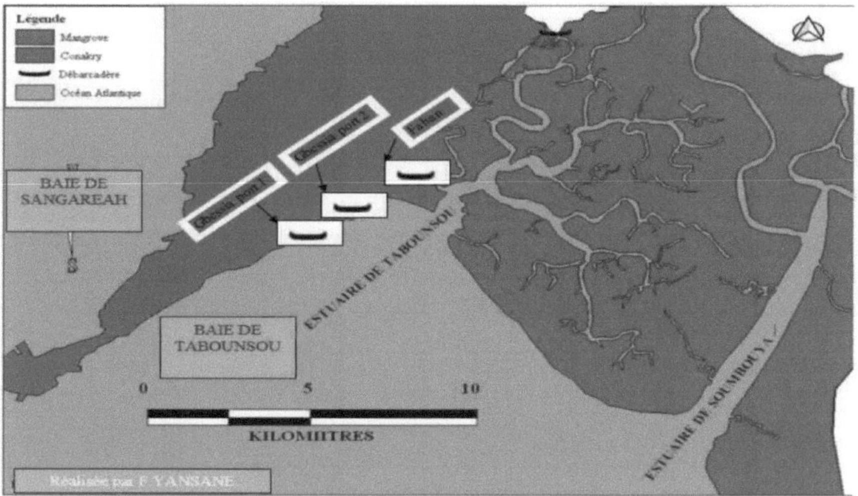

Figure II.2: Map of Tabounsou Bay

II.2.2. Socio-economic activities

The main activities carried out in Tabounsou Bay by the communities of Matoto and Coyah are essentially the same as those in the city of Conakry.

Table II.1: Equipment used

Hardware	Quantity
Animal material (Fish)	**60 individuals**
Field equipment	
Boat	1
GPS (Garmin)	1
Life jackets	1
Survey form	**42**
Riban	1
Block	1
Cooler	2
Identification key	2
Laboratory equipment	
Distilled water (litre)	1
Petri dishes	**61**
Ichtyometer	1
Graduated pipettes	2
Gloves (pack)	1
Precision scales	1
pH meter (Hanna)	1
Conductivity meter	1
Stomacher/Grinder	1
Scissors	1
Knife	1
Turbidimeter	1
Uviline spectrophotometry	1
Alcohol 70% (ml)	**500**
Reagents	
Zinc reagent 1 1RZ011	Zinc reagent 1 1RZ011
Copper Reagent 1 1RC060	Copper Reagent 1 1RC060
Lead pill reagent 1 1RN011	Lead pill reagent 1 1RN011
Cadmium Test Reagent 1 1RN011	Cadmium Test Reagent 1 1RN011

B. METHOD

The aim of this study was to assess the level of contamination of some fish species by trace metal elements (TMEs) in Tabounsou Bay with a view to proposing mitigation measures for these elements.

To achieve our goals, we adopted the following methodological approach:

1. Consultation of frameworks and analysis of archives ;

2. On-site survey ;

3. Sampling ;

4. Determination of the physico-chemical parameters of the water ;

5. Determination of metal levels in the flesh of fish species and in water ;

6. Proposed mitigation measures.

1. Consultation of frameworks and analysis of archives

Consultations were carried out through group and individual interviews with managers at the Centre de Recherche Scientifique Conakry Rogbanè (CERESCOR) and the Centre National des Sciences Halieutiques de Boussoura (CNSHB), in order to highlight the dangers of trace metals for fish and wildlife.

2. On-site survey

In the field, stakeholders (port authorities, fishermen and local residents) at the sites were interviewed individually and direct observations were made to gather information on environmental problems using survey forms.

3. Sampling

Sampling was targeted and focused on three elements (water sampling station, biological material and metals to be analysed). During this stage, water and fish samples were packaged and transported to the Office Nationale de Contrôle de Qualité (ONCQ) laboratory for analysis.

^ Sampling stations

We divided our area of investigation into three sampling sites, taking into account the level of exploitation of the environments: one site where human activity is very intense (**Faban**), a second less exploited site (**Gbessia port 2**) and a third site where human activity is intense (**Gbessia port 1**). Prior to sampling, new plastic bottles were rinsed twice with water from the site to be sampled. We took two samples of 0.5 litres of water per site at depths greater than 15 cm, in counter-current with the surface water, in January from 9 a.m. to 12 noon.

^ Biological material

The species ***Pomadasys jubelini, Mugil cephalus and Dentex angolensis*** were chosen for our study because of their nutritional value, their status as highly edible food for many local populations, their abundance (catches) in the study area and our financial resources.

In this work, we focused on the flesh of these species (the part that is generally consumed by humans).

^ Metals to be analysed

Our study focused on four metallic elements: Zinc, Copper, Lead and Cadmium. This choice was motivated not only by the availability of the reagents in the laboratory, but also by their great persistence in the environment, their ability to accumulate in the tissues of living organisms and to spread along the trophic chain, as well as their potential toxicity for ecosystems and human health, which is a global concern, and our financial resources.

4. Determination of physico-chemical water parameters

For the environmental analyses, the general parameters six parameters (temperature, hydrogen potential (pH), conductivity, salinity, turbidity and dissolved oxygen) were determined at the Physicochemical Laboratory of the National Office of Control and Quality then we compared these results with those of the standard admitted by the CE/2001 and other authors.

In order to avoid erroneous results, we rinsed the probes with distilled water or site water after each measurement and for each sample for all the physicochemical parameters measured.

- **To determine the temperature**

To determine the temperature, we used a device called a pH meter linked to a probe called a conductivity meter. We took 100 ml of water from the sample and put it in a 100 ml Erlenmeyer flask, then introduced the conductivity probe into the sample to be analysed and waited for the result of the water temperature on the screen.

- **To determine the pH (hydrogen potential)**

To determine the pH, the first step was to rinse the petri dish and the p^H -meter electrodes. This operation was repeated several times with distilled water and a quantity of the water sample to be analysed. The sample to be analysed (100 ml) was poured into the petri dish, then the electrode of the p^H -meter (p^H -meter Hanna) was immersed in the sample and waited until the value stabilised before reading the value on the screen.

The result displayed is recorded on pre-prepared cards and, at the end of the operation, the electrodes are rinsed two or three times with distilled water to prevent degradation.

- **For Turbidity determination**

We used a turbidimeter to measure the turbidity of the water. We took 5ml of water and put it in a 5ml bottle, then introduced the bottle into the machine, switched on the apparatus and waited a minute for the result to be displayed. The result thus obtained was recorded on a pre-established sheet. All 3 samples were analysed in the same way, right down to the last sample.

- **To determine Conductivity**

Using a device called a conductivity meter with a probe, we determined the electrical conductivity of the water. We took 100 ml of water and put it in a 100 ml petri dish, then immersed the conductivity meter electrode in the sample, switched on the device and waited a minute for the result to be displayed. The result was then recorded on a note card.

- **To determine Salinity**

To determine salinity, the first step was to determine the electrical conductivity of the water, using the same procedures as for pH determination, but this time with the conductivity meter. The standard conductivity of the seawater was then determined. To do this, a 35%o salt water solution was prepared, which enabled the ratio **R** between the conductivity of the sample and the standard solution to be calculated using the formula **R**= ^| (Emmanuel P. D., 2003);

With **CE**= Conductivity of the sample and **CS**= Conductivity of the standard solution and R the **CE/CS** ratio. The relationship **S‰= - 0.08996 + 28 .29720R + 12.80832R2 - 10.67869R^3 + 5.98624R4 - 1.32311R^5** , was used to determine salinity.

- **To determine dissolved oxygen**

Along with pH values, dissolved oxygen concentrations are one of the most important water quality parameters for aquatic life.

Dissolved oxygen remains a fundamental parameter for the distribution of aquatic organisms. Its variation depends on exchanges with the surface, and therefore with the wind, but also on the intensity of photosynthetic activity.

It was determined automatically using a photometer.

5.   **Assessment of metal levels in water and in the muscle of fish species.**

The samples were analysed by Uviline spectrophotometry in the physical-chemical laboratory of the Office National de Contrôle et de Qualité, and the results were compared with those of the standard accepted by CE/2001 and FAO/WHO (1989) and other authors. The following methodology was followed

^ **For water samples**

- **Determination of zinc (Zn) content**

Reagents required

Zinc reagent 1 1RZ011

Zinc 2 reagent 1RZ012

Preparation time: ~ 2 min

Preparing the sample

This part consists of taking 10 ml of the water to be analysed and introducing it into the Erlenmeyer flask. We then added 5 drops of zinc 1 reagent, homogenised and waited 1 minute.

Then we added another 10 drops of zinc 2 reagent, homogenised and filled the measuring tank.

White and sample measurement

This part is carried out in the apparatus (Uviline spectrophotometry).

In Concentration mode, we have selected analysis **461 Zn: 0.05 - 4.00 mg/L**

We then filled a tank with the water to be analysed without reagent (blank tank) and placed it in the instrument, pressing the "zero" button.

We then removed the cuvette, placed the sample cuvette to be analysed in it and pressed "Start" to take the measurement.

The result is recorded directly on a pre-established form.

- **Determination of copper (Cu) content**

Reagents required

Copper Reagent 1 1RC060

Copper 2 reagent 1RC070

Preparation time: ~ 3.5 min

Preparing the sample

This part consists of taking 10 ml of the water to be analysed and introducing it into the Erlenmeyer flask, then adding 5 drops of copper 1 reagent and homogenising.

We then added a further 10 drops of zinc 2 reagent, homogenised and then filled the measuring vessel and waited 3 minutes.

White and sample measurement

In Concentration mode, we have selected analysis **150 Cu: 0.05 - 5.00 mg/L**

We then filled a tank with the water to be analysed without reagent (blank tank) and placed it in the instrument, pressing the "zero" button.

We then removed the cuvette, placed the sample cuvette to be analysed in it and pressed "Start" to take the measurement.

The result is recorded directly on a pre-established form.

- **Determination of lead (Pb) content**

Reagents required

Lead pill reagent 1 1RN011

Lead pill reagent 2 1RN012

Preparation time: ~ 4 min

Preparing the sample

In this section, we took 10 ml of the water to be analysed and placed it in the Erlenmeyer flask, then added 1 level measuring spoon of Lead Pill Reagent 1 and mixed well.

We then added 10 drops of Lead Pill 2 Reagent, homogenised it, filled the measuring vessel and waited 3 minutes.

White and sample measurement

In Concentration mode, we have selected analysis **411 Pb: 0.1- 0.10 mg/L**

We then filled a tank with the water to be analysed without reagent (blank tank) and placed it in the instrument, pressing the "zero" button.

We then removed the cuvette, placed the sample cuvette to be analysed in it and pressed "Start" to take the measurement.

The result is recorded directly on a pre-established form.

- **Determination of cadmium (Cd) content**

Reagents required

Cadmium Test Reagent 1 1RN011

Cadmium 2 Test Reagent 1RN012

Preparation time: ~ 6 min

Preparing the sample

In this section, we took 10 ml of the water to be analysed and placed it in the Erlenmeyer flask, then added 7 drops of Cadmium Test Reagent 1 and mixed well.

We then added 7 drops of Cadmium 2 Test Reagent, homogenised it, filled the measuring tank and waited 3 minutes.

White and sample measurement

In Concentration mode, we have selected analysis **118 Cd: 0.001- 0.10 mg/L**

We then filled a tank with the water to be analysed without reagent (blank tank) and placed it in the instrument, pressing the "zero" button.

We then removed the cuvette, placed the sample cuvette to be analysed in it and pressed "Start" to take the measurement.

The result is recorded directly on a pre-established form.

- ^ **For the flesh of poisons** (same reagents for water)

- **Determination of zinc (Zn) content**

Preparation time: ~ 6 min

Preparing the sample

We took 10g of fish flesh in 90ml of distilled water (dissection and weighing) and then ground it in the Stomacher for 3 minutes;

We then took 10 ml of the grinding agent and introduced it into the Erlenmeyer flask (petri dish);

Then we added the necessary reagents and homogenised (stirring);

And wait a few minutes (about minutes) to rest.

White and sample measurement

This part is carried out in the apparatus (Uviline spectrophotometry).

In Concentration mode, we have selected analysis **462 Zn: 10 - 105 mg/L**

We then filled a tank with the water to be analysed without reagent (blank tank) and placed it in the instrument, pressing the "zero" button.

We then removed the cuvette, placed the sample cuvette to be analysed in it and pressed "Start" to take the measurement. The result obtained is recorded directly on a pre-defined form.

- **Determination of copper (Cu) content**

Preparation time: ~ 9 min

Preparing the sample

We also took 10g of the fish flesh in 90ml of distilled water (dissection and weighing) and then ground it in the Stomacher for 3 minutes;

We then took 10 ml of the grinding agent and introduced it into the Erlenmeyer flask (petri dish);

Then we added the necessary reagents and homogenised (stirring);

And wait a few minutes (about 5 minutes) to rest.

White and sample measurement

In Concentration mode, we have selected analysis **150 Cu: 10 - 110 mg/L**

We then filled a cuvette with the water to be analysed without reagent (blank cuvette) and placed it in the instrument, pressing the "zero" button.

We then removed the cuvette, placed the sample cuvette to be analysed in it and pressed "Start" to take the measurement. The result obtained is recorded directly on a pre-defined form.

- **Determining lead content**

Preparation time: ~ 11min

Preparing the sample

We also took 10g of fish flesh in 90ml of distilled water (dissection and weighing) and then ground it in the Stomacher for 3 minutes;

We then took 10 ml of the grinding agent and introduced it into the Erlenmeyer flask (petri dish);

Then we added the necessary reagents and homogenised (stirring);

And wait a few minutes (about 7 minutes) to rest.

White and sample measurement

This part is carried out in the apparatus (Uviline spectrophotometry).

In Concentration mode, we have selected analysis **412 Pb: 0.1- 0.10 mg/L**

We then filled a tank with the water to be analysed without reagent (blank tank) and placed it in the instrument, pressing the "zero" button.

We then removed the cuvette, placed the sample cuvette to be analysed in it and pressed "Start" to take the measurement. The result obtained is recorded directly on a pre-established form.

- Determination of cadmium content

Preparation time: ~ 15min

Preparing the sample

We took 10g of fish flesh like the others in 90ml of distilled water (dissection and weighing) and then ground them in the Stomacher for 3 minutes;

We then took 10 ml of the grinding agent and introduced it into the Erlenmeyer flask (petri dish);

Then we added the necessary reagents and homogenised (stirring);

And wait a few minutes (about 10 minutes) to rest.

White and sample measurement

This part is carried out in the apparatus (Uviline spectrophotometry).

In Concentration mode, we have selected analysis **118 Cd: 0.001- 0.10 mg/L**

We then filled a cuvette with the water to be analysed without reagent (blank cuvette) and placed it in the instrument, pressing the "zero" button.

We then removed the cuvette, placed the sample cuvette to be analysed in it and pressed "Start" to take the measurement. The result obtained is recorded directly on a pre-defined form.

6. Proposed mitigation measures

In this section, appropriate measures have been taken to mitigate the impact of discharges of industrial wastewater, industrial effluents, mining effluents, agrochemical discharges, etc.

CHAPTER III: RESULTS AND DISCUSSION

1. Consultation of frameworks and analysis of archives

Our discussions with executives from CERESCOR and CNSHB showed us that the problem of degradation of the Guinean coastal zone is a fact of life. It emerged from these consultations that more than 25% of the population of the southern coast of Conakry are involved in agriculture, and the main products used have an impact on aquatic animals.

Analysis of the archives has revealed an increase not only in the population but also in the amount of unhealthy waste produced in Conakry in recent years. Evaluation campaigns carried out by the Institut National de la Statistique **(INS)** show an increase in the population and the waste produced since 2000.

The figures below summarise the results of the archives

Figure III.1: Population and waste trends in Conakry

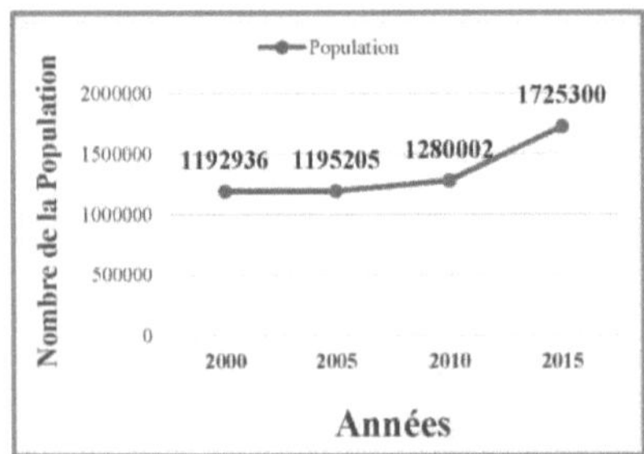

Figure III.1: Population growth in Conakry from 2000 to 2015

Figure III.2: Waste production by population from 2000 to 2015

Source: National Institute of Statistics **(INS)**

From these graphs, we can see that the number of people living in Conakry increased between 2000 and 2015, as did the amount of insalubrity.

In addition, the increase in insalubrity could be linked to the intensive urbanisation of the city of Conakry; we can deduce that as the population increases, so does the accumulation of waste. The transfer of this waste is essentially due to the leaching of soil and run-off water, which drains large quantities of products and residues into aquatic areas, with drastic consequences for species as a result of an excessive desire on the part of populations to move to the seaside for aeration in quantity and quality, rural exodus or emigration, migration and a high birth rate.

2. Field survey

In the investigation area, (5) ports were identified, the results of which are shown in the table below

Table III.1: List of ports surveyed and their main activities

N°	Sites	Number of people surveyed	Main activity	Species landed (names in Sousou)
1	Faban	12	Fishing, Carpentry, Agriculture and Fish Smoking.	Kèssi-kèssi, Fagba, Sori, Sinapa
2	Kinssy port	8	Cutting and sale of mangrove wood, fishing, agriculture and salt farming.	Bobo, Kouta, Sinapa
3	Landing Yimbaya Tannery	5	Fishing, smoking fish and cutting mangrove wood.	Fouta, Konkoé, Bobo
4	Gbessia port 1	10	Fishing, oyster harvesting, farming; carpentry and smoking.	Kèssi-kèssi, Fagba, Sori, Sinapa
5	Gbessia port 2	7	Fishing and agriculture.	Shellfish

From this table, we can see that a number of activities are carried out in the various localities in Tabounsou Bay: of the 42 people surveyed, 19 were fishermen; 11 were local residents (farmers); 5 were carpenters and 7 were woodcutters. These activities, which are the main ones carried out in the bay by local communities, especially farming and carpentry, cause a great deal of damage to the marine environment, particularly the degradation of the coastline following the use of biocides and the production of waste by carpenters (sawdust and planks).

This waste will be carried directly into the waters where the species live as a result of erosion, which will continue to pollute the environment and lead to changes in the structure and functioning of the ecosystem.

These results are in line with those found by (Konaté, S et *al.,* 2007), which indicate that of all the human activities carried out in Tabounsou Bay, agriculture, particularly rice growing, has the greatest impact on the bay and its resources. Clearing mangrove swamps for rice cultivation that is not sustainable and productive is a real waste of land and resources.

Table III.2: List and analysis of major problems in Tabounsou Bay

Affected elements	Major problems	Causes			Environmental impact	Socio-economic consequences
		Origin	Underlying	Immediate		
Fish and fisheries	Decline in fish stocks	Fishing agreements and illegal fishing.	Lack of fisheries management	Overfishing	Loss of biodiversity	Falling incomes and food insecurity
Coastal habitats	Physical alteration and destruction of coastal habitat (agriculture)	Anthropogenic pressures and natural factors (use of biocides)	Deforestation of mangroves and riverbanks	Erosion and sedimentation	Modification of the biotope	Migration, falling incomes, unemployment
Pollution and ecosystem health	Deterioration in water quality	Pollution from land-based activities and maritime transport	Insufficient environmental control resources	Domestic waste and industrial effluent.	Altered biodiversity Decline in biological productivity.	Economic loss; Human health problems.

The results of this table show that Conakry's southern coastline is facing serious threats, mostly due to poor human practices. As a result, the various resources are being severely disrupted or even destroyed, especially by pollution caused by waste dumped in the sea. This will cause fluctuations in the abiotic parameters of the species' habitat, a decline in fish stocks and a deterioration in water quality, all of which will have a dangerous effect on the health and well-being of the area's population as a result of the consumption of the products of these waters.

These ideas are in line with those of (Larno V., 2001), who conclude that most disturbances in the marine environment are often derived from human activities, such as agriculture (fertilisers, pesticides and agrochemicals), industry (trace elements and organic compounds), urban development (pathogens, organic substances, trace elements in waste water), tourism (litter, plastics on the coast) and shipping (oil spills).

Table III.3: List of fish landed in Tabounsou Bay from 2020 to 2021

N°	Family	Scientific names	French names (FAO)	Local names (Sousou)
1	Albulidae	*Albula vulpes*	Sea banana	Khomoukhomou
2	Ariidae	*Arius latiscutatus*	Gambian Machoiron	Konkoé
		Arius parkii	Machoiron	Konkoé
		Arius latiscutatus	Machoiron	Konkoé
3	Haemulidae	*Plectorhyncus macrolepis*	Large-lipped diagram	Kinsidinyi
		Pomadasys jubelini	Sompat grounder	Kèssi-kèssi
		Pomadasys peroteti		Kèssi-kèssi
4	Carangidae	*Caranx hippos*	Large trevally	Kawrè
		Caranx crysos	Common carangue	Kawrè
		Caranx senegallus	Senegal Caranx	Kawrè
5	Ephippidae	*Chaetodupterus gorensis*	Sea goat	Debelenyiforè
6	Claroteidae	*Chrysichthys nigrodigitatus*	Bagridcatfish	Khokhounyi
7	Cynoglossidae	*Cynoglosus monodi*	Sole language	Fagba
8	Sparidae	*Dentex angolensis*	Angolan tooth	Sinapa
9	Drepanidae	*Drepane africana*	African drepane	Débélenyi
10	Clupeidae	*Ethmalosa fimbriata*	African Ethmalosis	Bonga
		Ilishia africana	Razor	Laati
		Sardinella maderinsis	Flat sardinella	Bongasèri
11	Cichlidae	*Hemichromis fasciatus*	Carpe	Toka
		Tilapia guineensis	Carpe	Khobè
12	Thrichiuridae	*Trichiurus lepturus*	Sabrefish	Paniyèkhè
13	Sphyraenidae	*Sphyraena barracuda*	Barracuda	Kouta
		Sphyraena guachancho		Kouta

14	Mugilidae	*Liza falcipinis*	Large-finned mullet	Sèki
		Liza grandisquamis	Large-scaled mullet	Sèki
		Mugil cephalus	Cabo mullet	Sèki
15	Polynemidae	*Galeoides decadactylus*	Captain Plexiglace	Sanoussi
		Pentanemus quinquarius	Captain moustache	Gbalakassa
		Polydactylus quadrifilis	Big apitaine	Sori
16	Monodactylidae	*Monodactylus sebae*	Candlefish	-
17	Psettodidae	*Psettodes belcheri*		Fagbakhamè
18	Sciaenidae	*Pseudotolithus senegalensis*	Otolith bobo	Sosoékondounké
		Pseudotolithus brachignatus		Fouta
		Pseudotolithus elongatus	Hunchbacked otolith	Boboè
		Pseudotolithus epipercus		Boboè foret
19	Bothidae	*Syacium micrurum*	Fake dab paté	Fagba forè
20	Dasyatidae	*Dasyatis margarita*	Pearl stingray	Kouléyèkhè
21	Rhinobatidae	*Rhinobatos cemiculus*	Guitar ray	Matéki

Analysis of this table shows that the Faban estuary is very rich in fish resources, with 21 families of fish identified, divided into 38 species frequently landed, with *Ethmalosa fimbriata* predominating.

These results differ from those of (Bah, T.H., 2016), which recorded 46 species of fish in the same estuary, with *Ethmalosa fimbriata* being the most abundant species caught. This would be due to the fact that these species move in shoals, which makes them more vulnerable to certain fishing gear and techniques, or to their high reproductive capacity.

These results show a clear reduction in the number of species present in the area, which is thought to be due to overexploitation and heavy water pollution, which has led to the migration of species, as the Faban area is one where human pollution is very high as a result of heavy urbanisation. The abundance of this species in the catches is thought to be due to the fact that it can be fished using several gears (hawksbill, encircling gillnets, drifting gillnets, etc.).The abundance of this species in the catches would be due to the fact that it is fished by several gears (sparrowhawks, encircling gillnets, drifting gillnets, etc.) but also to the fact that it reproduces all year round compared with the other species recorded, such as *Trichiurus lepturus*, which produce many eggs.

Ethmalosafimbriata, for example, is a species that can lay more than 2,000,000 (two million) eggs in a single clutch. *Trichiurus lepturus,* on the other hand, is a less productive species, as not only does it grow very slowly, but the female reaches her first sexual maturity at a size of 69.3 cm (FAO, 2008).

3. Analysis of water physico-chemical parameters

In order to determine the physico-chemical qualities of the water, an analysis was carried out in the laboratory, and the results are shown in the following figures:

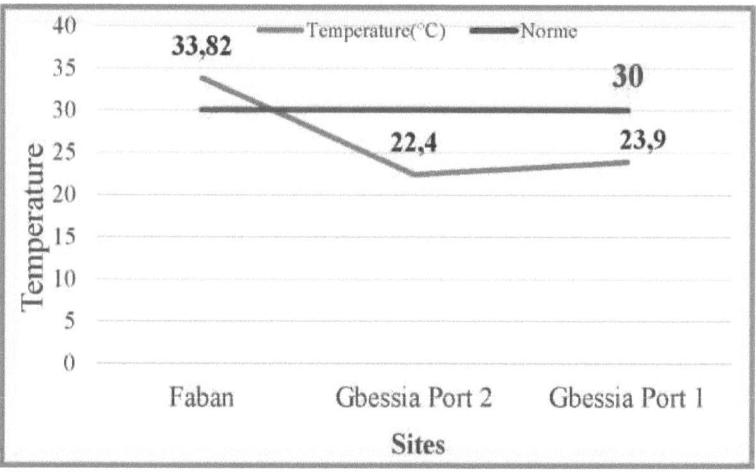

Figure III.3 : Temperature variation

This figure shows a sharp rise in temperature in the Faban area, slightly higher than those reported by (Bah A.,2019) in the same bay at Sangoyah (32.91°C) and by (Onivogui G et *al.*,2013) in the Konkouré estuary (31.02°C). This difference in temperature in these areas is thought to be linked to the large volume of wastewater, which is generally warmer and causes the water temperature to rise.

On the other hand, they are consistent with the results of the work of (Pezennec, O., 1999) who indicates that hydrologically, Guinean waters remain warm throughout the year and are close to the ambient temperature of the area, which is the result of its tropical climate.

We also note the variation in temperature in the Gbessia port 2 and Gbessia port 1 zones, with values of 22.4°C and 23.9°C, which are below the WHO standard of 30°C.

These sharp variations in temperature in these areas are the result of heavy pollution of these waters, which will have a serious impact on the life of species, as temperature is the driving force behind all the biological functions of species (reproduction, feeding, growth) and the abiotic factors of an ecosystem.

According to (Rodier J., 1984), temperature is an ecological factor that conditions the distribution of aquatic organisms. It is of vital importance directly in the metabolic activity of organisms, or indirectly by modifying the ecological factors of the environment and consequently their

répartition biogéographique.

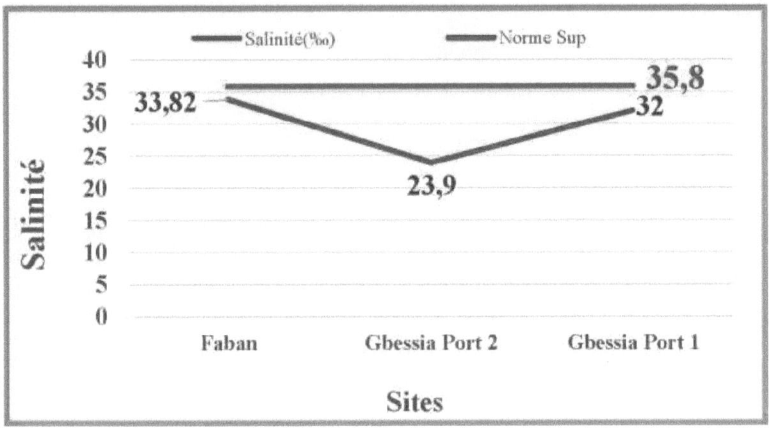

Figure III.4: Variation in salinity

The figure shows that haline values are below the WHO standard of 35.8%.

Our results remain comparable to those reported by (Bah A.,2019) at Boulbinet (33.7%) and by (Bangoura K et a/.,2013 at Kakounssou in the Konkouré estuary (33.04%).

These variations in salinity at these sites may be due to variations in temperature, rainfall, turbidity of the environment as a result of excessive waste, which will have an impact on the life of the species, as the lower the salinity, the less favourable the environment becomes for stenohaline species.

These results are consistent with those found by (Pezennec, O., 1999) who maintains that in Guinea, salinity varies according to the season (ranging from 2% in the rainy season: July-August, at the mouths to 34 g/l in the dry season).

According to (Tamoïkine, M.Y., 1994), in the Guinean estuarine environment the salinity level is always below 35 g/l and varies from 33 g/l to 0 g/l within about 20 km of the mouths.

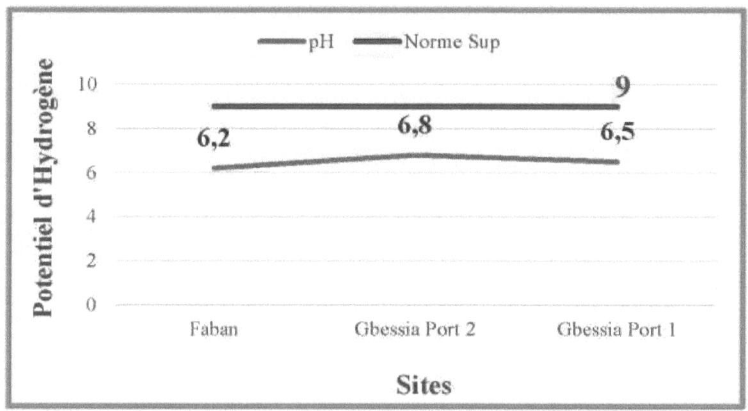

Figure III.5 : Variation in pH

This figure shows that the pH of the water is between 6.5 and 8.5.

However, our pH values are relatively lower than those measured by (Bah A.,2019) at Boussoura (8.3) and by (Bangoura K et *al* 2013) at Bokhinènè (6.98).

These variations in hydrogen potential (pH) at these sites may be due to the input of organic matter drained by wastewater, which leads to a decrease in pH. These concentrations are confirmed by water turbidity data, which show significant values, particularly at Faban, and which would have reduced the photosynthetic activity of the algae and therefore the pH of the water.

These results are in line with those of (Derwich E., 2010) who point out that in natural waters, pH values are between 6 and 8.5 and decrease in the presence of high levels of organic matter and increase during periods of low water, when evaporation is high.

According to (Copin, M.G. 2002), pH has a direct effect on the availability of metal ions in the marine environment, especially when the medium is acidic (low pH), and on the rate at which they accumulate in organisms. In neutral or basic waters, they precipitate and accumulate mainly in the solid phase (sludge).

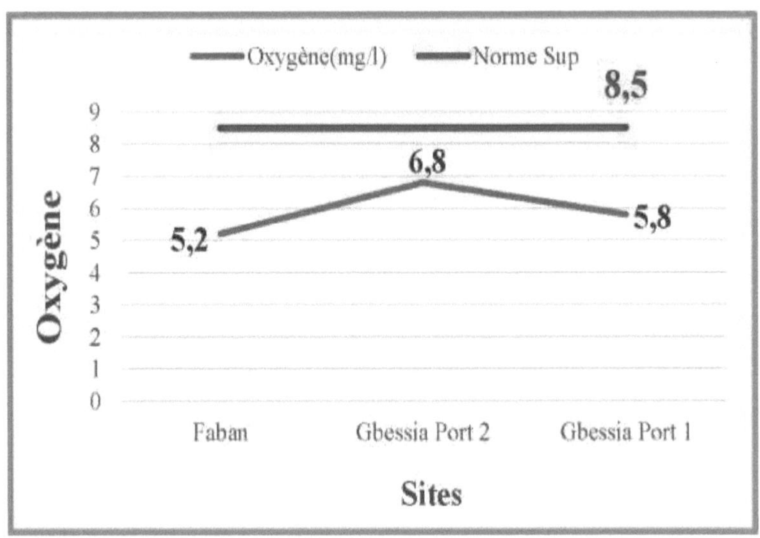

Figure III.6 : Variation in oxygen content

This figure shows a variation in dissolved oxygen content at Faban and Gbessia port 1 of 5.2 mg/l and 6.8 mg/l respectively.

Our results are consistent with those obtained by (Bah. A., 2019) at the Port Autonome de Conakry (5.6 mg/l) and by (Baldé S et *al.,* 2013) in the Konkouré estuary (6.5 mg/l).

These decreases in dissolved oxygen levels would, on the other hand, be linked to high turbidity, temperature, light penetration, water agitation, nutrient availability and hydrodynamic calm, which prevents water mixing.

This dissolved oxygen concentration is also a function of the rate at which the environment is depleted of oxygen by the activity of aquatic organisms and the processes of oxidation and decomposition of the organic matter present in the water.

Overall, the closer the dissolved oxygen (DO) concentration is to saturation, the greater the capacity of the sea or estuary to absorb pollution.

According to (Guisse, A., 2012), dissolved oxygen concentration is a fundamental state variable that is involved in many processes; it is also a good indicator of the health of an ecosystem.

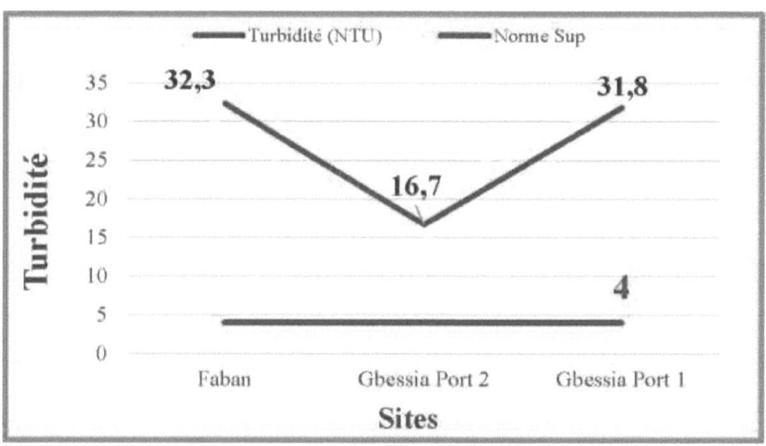

Figure III.7 : Turbidity variation

The highest turbidity value was observed at Faban (32.3 NTU) and the lowest at Gbessia port 2 (16.7 NTU).

However, these results are still higher than the values recorded by (Bah. A., 2019) at Yimbaya (14 NTU) and those found by (Barry M.K et *al* 2013) in the Konkouré estuary (1.86 NTU).

These high levels of turbidity can be explained by the fact that our sampling sites are close to landing stages where human activity is intense, but also by the mixing of salt water and run-off water, which is the cause of the high turbidity. When the water is turbid, aquatic plants receive less light.

Plant growth and photosynthesis - the production of oxygen by plants - are therefore greatly reduced. The aquatic food chain is affected.

Our results are in agreement with the work of (Diané, L. *et al.,* 2005) on Sangareah Bay, in which he maintains that the high turbidity of the water is recorded during the low-water period and can be explained by the large input of solid particles from the leaching of soil from the catchment caused by rainfall due to deforestation and soil degradation.

However, all these sites have a turbidity value much higher than that allowed by the WHO, which will cause significant damage to both organisms and water quality.

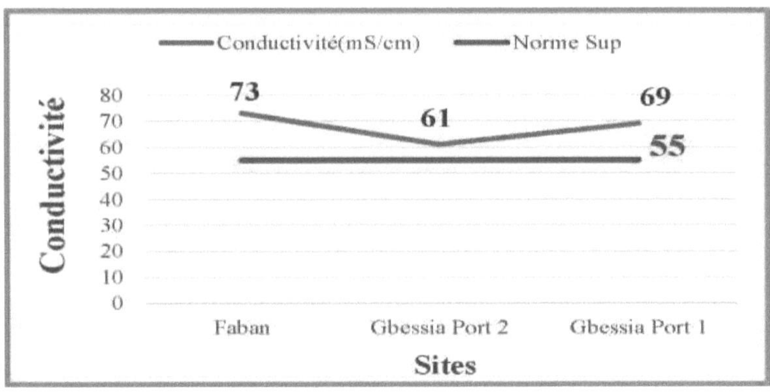

Figure III.8 : Variation in Conductivity

Based on these results and the values recorded at the various sites (61 mS/cm) at Gbessia port 2, (69 mS/cm) at Gbessia port 1 and (73 mS/cm) at Faban, the conductivity of the water is high compared with the WHO standard of 55 mS/cm.

Our degree of mineralisation turns out to be higher than that measured by (Bah. A., 2019) in the same bay (55.83 mS/cm) and by (Fouad S et *al* 2014) in Morocco (53 mS/cm).

This situation of high dissolved salt mineralisation is thought to be due to anthropogenic activities such as the direct discharge of: liquid waste from toilets, baths using antiseptics, washing powder, dishes and waste from factories processing agricultural products, cosmetics and others.

These results are equivalent to those of (Ouali N.,2018), in Algeria in the Bay of Oran, which indicate that the higher the conductivity of the water, the richer it is in mineral salts, and aquatic species generally cannot tolerate significant variations in dissolved salts.

4 Assessment of metal content in the water and in the muscles of the three species of fish To assess the content of metal ions present in the water, we carried out an assay of these ions, the results of which are shown in the following figures:

a) Assessment of metal content in water

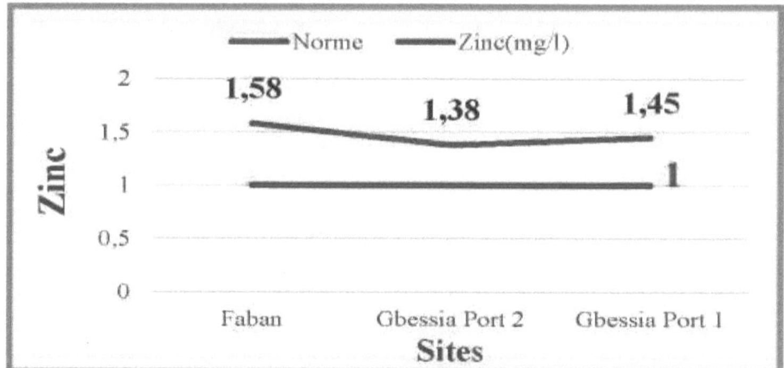

Figure III.9: Zinc concentration

From this figure, the different concentrations of Zinc found in the samples taken at the different sampling sites, it is easy to see that the highest concentration (**1.58mg/l**) was obtained in the samples taken at Faban.

These recorded values are within the Zinc range obtained by (Bah A.,2019) in Sonfonia and Dabondy (0.23 mg/l and 2.48 mg/l) and that reported by (Bangoura K et *al*.,2013) in the Konkouré area (0.1 mg/l and 180 mg/l).

This increase could be due to a significant input of agricultural manure (animal feed, slurry) and urban activities (road traffic, waste incineration) to these sites.

In all cases, our recorded Zinc values at the sites, 1.58 mg/l, 1.38 mg/l and 1.45 mg/l respectively, are below the World Bank standard of 1 mg/l.

According to Rodier P. (1984), Zinc has a certain toxicity for aquatic life. This toxicity depends on the degree of mineralisation of the water and the area under consideration. It affects fish from a few milligrams per litre and is also influenced by the hardness of the water, its oxygen content and the temperature.

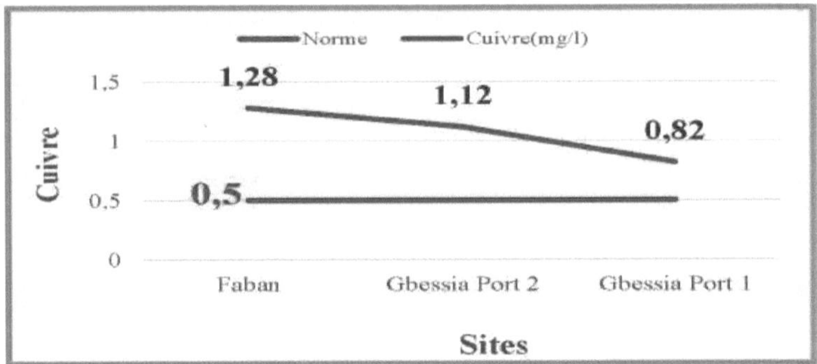

Figure III.10: Copper concentration

From this figure, we can see that compared with the discharge standard accepted by the World Bank (0.5 mg/l), there is a high concentration of the metal copper in these waters, at 1.28 mg/l, 1.12 mg/l and 0.82 mg/l respectively.

However, our copper values are relatively lower than those measured by (Bah A., 2019) at the Centrale Thermique du Kaloum (CTK), which receives permanent wastewater and water from natural sources, and higher than those reported by (Barry M.K et *al.*, 2013) in the Konkouré (Kakounssou) area.

These variations in copper content at these sites may be linked to terrigenous inputs and various industrial activities such as the Alpha soap factory, the Flour factory, the Capri-Sun factory, etc. However, these values are still below the limit values set by the World Bank for discharge (0.5 mg/l).

According to (Marchand M et *al.*, 1997), a copper concentration ranging from 0.1 to 1mg/l is not dangerous for fish, but the same concentrations are dangerous for certain aquatic species (molluscs) and the toxicity of copper varies depending on the species and the physico-chemical characteristics of the water (temperature, hardness and quantity of free CO_2).

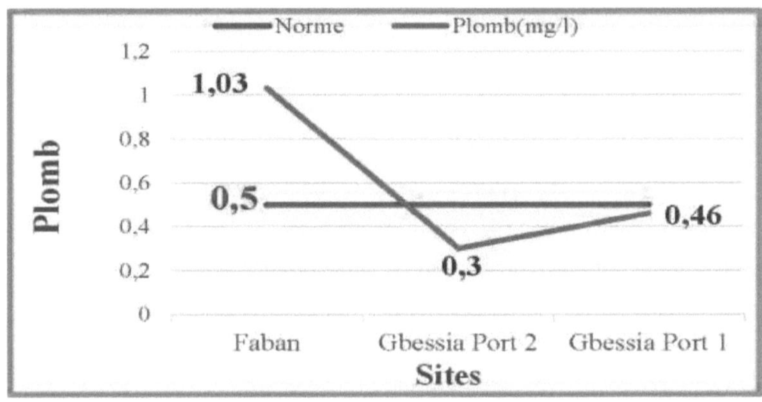

Figure III.11 : Lead concentration

On this figure, compared with the standard accepted by the World Bank (0.5 mg/l), we note that at the Gbessia port 2 and Gbessia port 1 sites the values found are lower than the discharge standard (0.3 mg/l and 0.46 mg/l), which indicates that these sites are not very polluted with this substance.

At the Faban site, on the other hand, this limit is significantly exceeded (1.03 mg/l compared with 0.5 mg/l). This indicates a risk of pollution by this metal through the effluents discharged at this site. Our results are comparable to those reported by (Bah A., 2019) at Boulbinet (0.91 mg/l) and by (Baldé S et al., 2013) in the Konkouré estuary (1 mg/l).

The variation in these values at the various sampling sites can be explained not only by diffuse pollution (discharges and inputs due to the existence of industrial sites), indirect inputs through road washout and rainwater, but also by their proximity to a major road, as this element is used as an anti-knock agent in petrol.

According to (Yassine M., 2011) the micro-organisms responsible for the aerobic degradation of organic matter are sensitive to lead from 0.1 mg/l. He also states that lead toxicity affects plants as well as animals and humans. Copper toxicity in the aquatic environment is highly dependent on alkalinity, pH and the presence of organic matter.

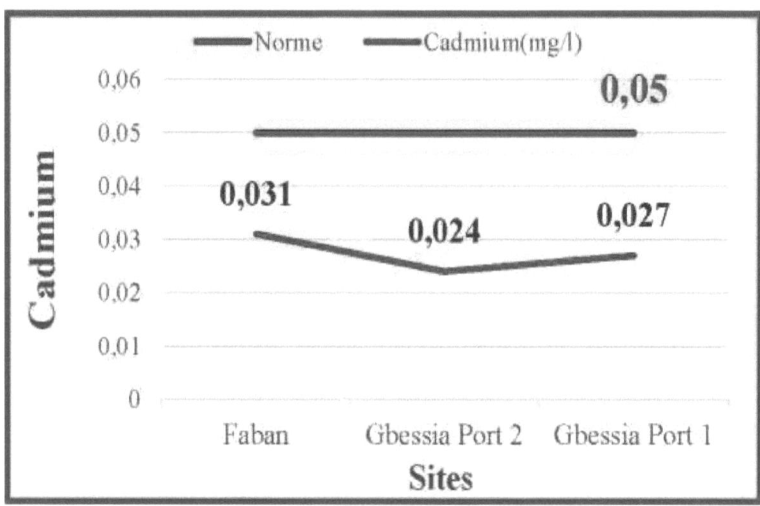

Figure III.12: Cadmium concentration

In general, this figure shows that all the samples from the 3 sampling sites analysed contain this metal at different concentrations, but compared with the accepted standard for this metal (0.05 mg/l), none of the concentrations exceeds the World Bank standard;

This indicates that these environments are less polluted by this metal because it is less present in the water discharged by third parties, which is a considerable advantage not only for the fish but also for the quality of the water and consumers, as it is a heavy metal.

This is confirmed by (Bah A., 2019) in Sangareah Bay at Sonfonia and by (Onivogui G et *al.,* 2013) in the Konkouré estuary, which report similar cadmium levels to ours (0.025 mg/l and 0.010 mg/l).

The presence of this metal at these various sampling sites may be linked to discharges from chemical textile and dyeing industries that contain this element, such as SOBRAGUI, SOGUIPAH, TOPAZ MULTI INSDUSTIE, etc.

b) Assessment of average metal content in the three species

In order to evaluate the content of metal ions present in certain species, we carried out an assay of these ions in three species of fish. The results we obtained are shown in the following figures:

^ *Pomadasys jubelini*

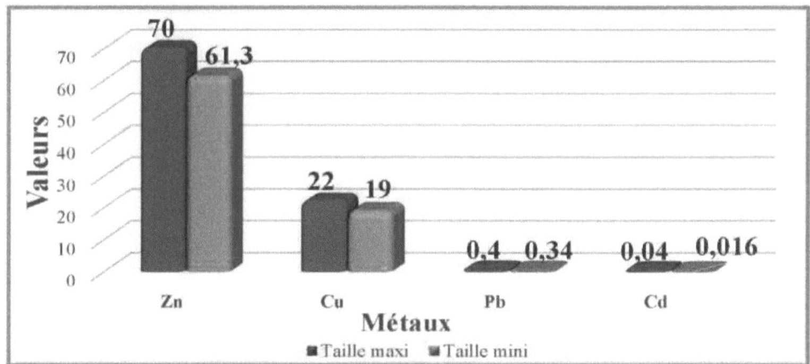

Figure III.13: Evaluation of average metal contents (Cd, Cu, Pb, Zn) in mg /l
in *Pomadasys jubelini.*

From this figure, the concentrations of Cadmium and Lead are less significant than those of Copper and Zinc. Cadmium concentrations are very insignificant compared with the other metals, but Zinc concentrations are very high with an average of 70 mg/l (maximum size) and 61.3 mg/l (minimum size). Our values are higher than those found by (Canli G et *al.*, 2002) in Turkey (34.58 mg/l) and lower than those reported by (Ouali N., 2018) in Annaba Bay in Algeria (82 mg/l) for the same species.

These variations in the rate of zinc accumulation by this species *(Pomadasys jubelini)* in these areas may be linked to the input of domestic wastewater containing this element. Although zinc is an essential metal for the metabolism of aquatic organisms, its significant accumulation can damage their health.

With reference to the European guide values (EC/2001 Recommendations) and those of FAO/WHO 1989 on the edibility of fish flesh, none of the values exceed the required standards (100 mg/l).

^ *Mugil cephalus*

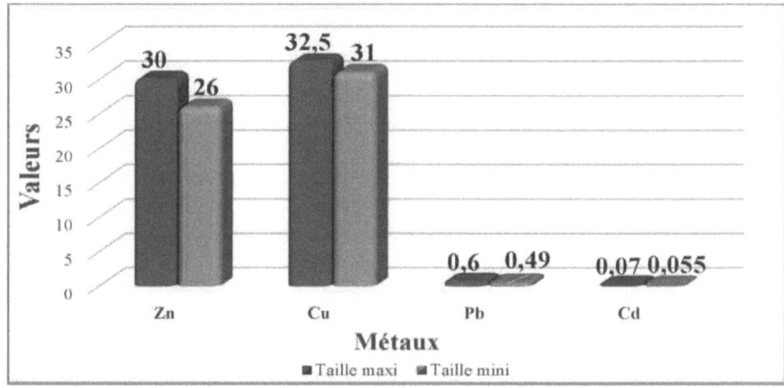

Figure III.14: Evaluation of average metal content (Cd, Cu, Pb, Zn) in mg/kg in *Mugil cephalus*

The results in this figure show variable doses of the four heavy metals (Zn, Cu, Cd, Pb) in the flesh of the individuals (**mullet)** in the two batches studied; our data clearly indicate that the average concentrations of Copper found in the two batches are higher than those of the other three metals.

With regard to the FAO 1989 limit values and those of the EC 2002, on the edibility of fish muscles, copper levels are higher (32 mg/l) than the standard required for this species (30 mg/l). Our copper values are comparable to those reported by (Ben Mansour P., 2009) in Algeria (33 mg/l) and by (Dural et *al.*, 2009) in Turkey in Karatas Bay (31 mg/l) with the same species.

The high level of copper in this species could be linked not only to its mobility and food preference, but also to the fact that copper is an essential metal for the metabolism of aquatic organisms; it therefore accumulates quite significantly in this species.

In view of these results, it remains to be seen whether the consumption of *Mugil cephalus* from Tabounsou Bay is dangerous in terms of the toxicological risks that its contamination may entail.

As a result, the consumption of **mullet** from the bay remains a formidable and dangerous option in terms of the toxicological risks that contamination may entail.

^ *Dentex angolensis*

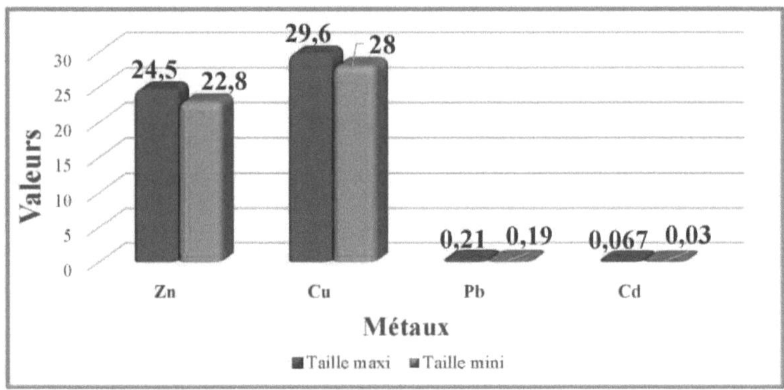

Figure III.15: Evaluation of average metal content (Zn, Cu, Pb, Cd) in mg/kg in *Dentex angolensis*

The results reported in this figure clearly indicate that the average concentrations of Copper and Zinc found in the flesh of ***Dentex angolensis*** are higher than those of Cadmium and Lead in the various batches, whatever the size of the specimens considered. This indicates that these two metals are more present in these waters as a result of human (industrial) and natural inputs and have a high contamination capacity.

We note that all the average concentrations of these metallic trace elements in the flesh of ***Dentex angolensis*** fish in Tabounsou Bay are below the standard accepted by the FAO/WHO and that of the European Community EC 2002. Our results obtained for Zinc and Copper levels are higher than those reported by Sahbaoui (2015) who worked on the same species at Ghazaouet in Algeria (25 mg/l and 28.5 mg/l). Our values obtained for the average Lead and Cadmium contents are also comparable to those reported by (Goual., 2000) who had worked on the same species at Beni-Saf in Algeria.

These variations in the rate of accumulation of these metals by this species may be due to diffuse pollution at these sites.

5. Proposed mitigation measures

Following our findings and in the light of the results obtained during our investigations, a set of corrective measures has been proposed to reduce the effects of trace metals on the population in general and on fishery resources in particular:

To the State

- ✓ Consider setting up a surveillance network along the Guinean coast;

- ✓ Create wastewater treatment plants in major cities;

- ✓ Encourage factories to set up effluent treatment plants by implementing a good policy;

- ✓ Applying current regulations on environmental degradation through monitoring, controls and surveillance.

Political decision-makers, local authorities and local residents

- ✓ To take appropriate measures to mitigate the impact of discharges of industrial wastewater, mining effluents, agrochemicals, etc. ;

- ✓ Raising awareness among the general public by keeping them informed of the dangers and risks to their food safety;

- ✓ Develop the coastal area and purify the water ;

- ✓ Encourage farmers to reduce the use of biocides, including persistent organic pollutants;

- ✓ Set up a body to control and monitor the quality of industrial and domestic effluent (equipped with appropriate resources).

To future researchers

The preliminary results obtained from this study open up recommendations for future perspectives that would be judicious to undertake:

- ✓ Continue this study over several cycles in order to gather as much information as possible on the levels of TME contamination in sediments and living organisms;

- ✓ The next logical step is to extend this work to the entire Guinean coastline;

- ✓ The list of trace metal elements needs to be extended to include other contaminants;

- ✓ We also need to understand the impact of this contamination on organisms on an individual scale and possibly on a population scale. This type of study can only be envisaged through a multidisciplinary approach combining chemistry, biology and ecology.

CONCLUSION

At the end of this study on the theme of **"Studies of metallic pollution on some species of fish caught in the Tabounsou-Conakry bay"**, the results obtained on the state of water quality at the various sampling sites seem to highlight the direct impact of global pollution, generated by the discharge of wastewater from these localities.

The physico-chemical quality of the bay's water reveals :

-High mineralisation of the water, as indicated by high temperature and electrical conductivity values. This mineralisation is not only a natural phenomenon resulting solely from the geological context of the region, but also from the input of wastewater from the southern part of Conakry.

-Significant pollution, confirmed by the high levels of Zinc and Copper. This pollution could be exacerbated by surface leaching from fertiliser-treated agricultural soils, as well as industrial and domestic discharges from units and dwellings along the shores of the bay.

According to the average metal contents recorded in our study, we can observe :

- ✓ Definite contamination of the study area, revealed by the presence of heavy metals (Zn, Cu, Cd, Pb) found in the flesh of these three species of fish and in the water;

- ✓ The results of this study showed that these species accumulate trace elements to varying degrees;

- ✓ The accumulation gradient for these species is as follows: Zn > Cu > Pb > Cd ;

- ✓ The metal concentrations analysed in the flesh of *Mugil cephalus,* copper in the first **batch**, showed values higher than the standards established by FAO/WHO.

At the end of this work and in view of the results obtained, *Mugil cephalus* could constitute a highly satisfactory sentinel organism in the context of a multi-compartment approach to monitoring contamination in the marine environment.

In short, the presence of these metal ions in these waters is clear proof that they are polluted as a result of human activity, precipitation and erosion, posing a real threat not only to the health of the organisms that live there but also to the consumers of these organisms that we are.

To deal with this very worrying situation, it is more than urgent to take measures such as

In relation to the State

- ✓ Move rubbish dumps away from shorelines, seas and estuaries;

- ✓ Raising public awareness of the need to adopt responsible behaviour in relation to watercourses;

To the authorities of the Institut Supérieur des Sciences et Médecines Vétérinaires (ISSMV) in Dalaba

- ✓ Strengthen the capacity of the ISSMV/D laboratory to meet the equipment needs of the Fisheries and Aquaculture Department.

BIBLIOGRAPHICAL REFERENCES

1. Abdoulaye Bah 2019: Impacts of effluents from Conakry's main thermal power plants on the marine and coastal environment.

2. Amiard, J.C., Amiard-Triquet, C., Berthet, B., Metayer, C. (2008). Contribution to ecotoxicoloical study of Cadmium, Lead, Copper and Zinc in the mussel Mytilus edulis. Mar Biol, (90). pp- 425431.

3. Bauchot, M.-L. and J.-C. Hureau, 1990. Sparidae. In J.C. Quero, J.C. Hureau, C. Karrer, A. Post and L. Saldanha (eds.) Checklist of the fishes of the eastern tropical Atlantic (CLOFETA). JNICT, Lisbon; SEI, Paris; and UNESCO, Paris. Vol. 2. 790-812

4. Berg D.M., Raven.P.H Et Hassen Zahl D M , 2009-Boech Environment 6th edition

5. Bester, C. (2004). "Ichthyology at the Florida Museum of Natural History" (On-line). Accessed October 16, 2005.

6. Bouchriti, 2003- surveillance des zones de production conchyticoles actes du séminaire sur la qualité des produits de la pêche ,20-24 mai 2002, casablanca,maroc,édité par insamak.

7. Boutiba, Z,2004- Guid de l'environnement marin Edit: DAR EL GHARB,273 P.

8. Casas, S, 2005- modelling of the bioaccumulation of trace metals (Hg,Cd,Pb,Cu and Zn) in the mussel Mytillusgalloprovincidis in a Mediterranean environment, doctoral thesis ENV MAR univ sud Toulon ,314P.

9. Chabanne (J), 1987 - Le peuplement des fonds durs et sableux du plateau continental sénégambien, Etude de sa pêcherie chalutière, biologie et dynamique d'une espèce caractéristique le rouget (Pseudupeneus prayerisis). Paris, ORSTOM, *Etudes et Thèses,* 355 p.

10. Champeau, O. (2005). Biomarkers of effects in C. Fluminea: from laboratory development to mesocosm application. Doctoral dissertation, Bordeaux 1 (France).

11. Chiffoleau J F., Claisse, D., Cossa, D., Ficht,A.,Gonzalez,J. , Guyot T ,Michel, P.,Miramand,P;Oger,C Et Petit. 2001- La contamination métallique, programme scientifique seine aval : 39p.

12. Chiffoleau, J.F., Auger D., Chartier, E., Michel, P. (2001). Spatiotemporal changes in Cadmium contamination in the Seine estuary (France). Estuaries 24 (6B): 1029-1040

13. Christian, N; Alain, R; 2004-Déchets et pollution, impact sur l'environnement et la santé, Dunod, paris.pp100.

14. Copin, M.G. (2002). Chemistry of seawater. Institut Océanographique. Synthesis collection.

15. Cravez V et Bernard G, 2006- Pollution marine : les définitions www.nuiv-mrs.fr. C.N.R.S., 2005- (centre nationale de la recherche scientifique) "Principaux rejets industriels".

16. Derwich E., Benaabidate L., Zian A. (2010), Journal 8 101-1

17. Desautels, M ; 2005-l'eutrophisation de nos plans d'eau, c'est quoi fiche technique n0 2 Quebec canada,2P. DESAUTELS, M; 2005-l'eutrophisation de nos plans d'eau, c'est quoi fiche technique n0 2 Quebec canada,2P.

18. Diané, I., 2005. Influence of human activities on the hydrodynamic functioning of estuarine ecosystems (abiotic factor): the case of Sangaréyah Bay.

19. Directive 2000/60/EC of the European Parliament and of the Council of 23 October 2000 establishing a framework for Community action in the field of water policy. Published online on 22 December 2000, consulted on 08 January 2009.

20. Directive 2000/60/EC of the European Parliament and of the Council of 23 October 2000 establishing a framework for Community action in the field of water policy. Published online on 22 December 2000, consulted on 08 January 2009.

21. FAO/WHO. (1989). Evaluation of certain food additives and the contaminants mercury, lead and cadmium, WHO Technical Report, Series No. 505.

22. Gagneux-Moreaux S., 2006 - Les métaux (Cd, Cu, Pb et Zn) dans la production des microalgues sur différents milieux de culture : biodisponibilité-bioaccumulation et impact physiologique. PhD thesis in marine biology. University of Nantes.257P

23. Gagneux-Moreaux S., 2006 - Les métaux (Cd, Cu, Pb et Zn) dans la production des microalgues sur différents milieux de culture : biodisponibilité-bioaccumulation et impact physiologique.Thèse de doctorat en biologie marine. University of Nantes.257P

24. Gbago ONIVOGUI ; Saidouba BALDE ; Kandet BANGOURA and Mamadou Kabirou BARRY Assessment of heavy metal pollution risks (Hg, Cd, Pb, Co, Ni, Zn) in water and sediments of the Konkouré river estuary (Rep. of Guinea)

25. GOEURY D. (2014). *Marine pollution.* Paris, France.

26. Guisse, Amed (2012). Influence of some physico-chemical parameters on the Horizontal Distribution of the Ichthyo-planktonic Fauna in the Sonfonia estuary. Chemistry and biology. Master's thesis. Conakry. Gamal Abdel Nasser University, Conakry. P4-11.

27. Hebbar, C, 2005-Surveillance de la qualité bactériologique des eaux de baignades cas des plages d'ainfranin et de Kristel, magister thesis, University of ORAN, 228P.

28. Hemalatha, S., Platel, K., Srinivasan, K. (2006). Zinc and iron contents and their bioaccessibility in cereals and pulses consumed in India. Food Chemistry, 102, 13281336.

29. Hemida F., Cherabi O., Nouar A. and F. Amire, 1995. Clé de détermination de la famille des Sparidés : proposition pour une démarche nouvelle. Actes du 1ère Congrès Maghrébin des Sciences de la Mer: 34p.

30. Hill, K. (2004). Smithsonian Marine Station at Fort Pierce. (On-line). Accessed October 16,2005 at http://www.sms.si.edu/irlspec/Mugil cephalus.

31. Khelil,F, 2007- évaluation de la contamination de l'eau de mer et d'un mollusque la moule Mytillusgalloprovincidis pêche du port d'Oran , mémoire de magistère , Univ d'oran,112P.

32. KHELIL,F, 2007- évaluation de la contamination de l'eau de mer et d'un mollusque la moule

Mytillusgalloprovincidis pêche du port d'Oran, mémoire de magistère, Univ d'oran,112P.

33. Konaté, S.; Keita, I. K.; Keita, A. et al]. (2007). Study and monitoring of the physicochemical, biological and sedimentological characteristics of Guinean coastal waters. CERESCOR report on the GCLME - ONUDI project N° GP/ RAF/ 04 / 004. p63.

34. Lafabrie, C. (2007). Utilisation de Posidonia oceanica (L.) Delile comme bio-indicateur de la contamination métallique (Doctoral dissertation, Université de Corse).

35. Lane, T. W. & Morel, F. M. (2000). A biological function for cadmium in marinediatoms. *Proceedings of the National Academy of Sciences, 97*(9), 4627-4631.

36. Larno, V., Loroche, J., Launey S., Flammaion P., Devaux A., 2001 - responces of chub (leucescuscephalus) populations chemical stress, assessed by genetic markeus.DUA damage and cytochrome P 4501A induction, Ecotoxicology 10:175

37. Leblanc.J.C. C.2004, étude de l'alimentation totale française. Mycotoxins, minerals and trace elements .INRA. Ministère de l'agriculture, de l'alimentation de la pêche et des affaires rurales .72P.

38. Lidsky T.I., Schneider J.S., 2003. Lead neurotoxicity in children: basic mechanisms and clinical correlates. Brain 100: 284-293.

39. Maadjou BAH 2015 report on the implementation of the marine and coastal biodiversity programme

40. MALQUIOT M et BERTOLINI P H, 2000-Encyclopédie de l'environnement et du développement durable,1100mots,Ed,Mecy consult,192P.

41. Mansouri Kahina & Khenache Lila 2015 Contribution to the study of the accumulation of heavy metals (Zn, Cu, Cd, Pb) in the muscle and visceral mass of *Pomadasys jubelini* caught in the Gulf of Bejaia

42. Marchand M. & Yasi *Chemical contaminants in aquatic environments.* Oceanis vol. 22 (2), 1995, vol. 22 (3), 1996, vol. 23 (4), 1997.

43. Miquel, M. (2001). The effects of heavy metals on the environment and health. Report Office Parlementaire d'évaluation des choix scientifiques et technologiques (Dir.).

44. Nakhlé, B. (2005). Numerical modelling of flood or submersion waves

45. NAS/NRC (1989). Recommended dietary allowances, National Academy of Science/National Research Council, Washington.

46. Nemery, F.J ; Mono, V ; Navratil, O (2010) : Feedback on the use of turbidity in mountain rivers; TSM. Technique sciences méthodes, génie urbain génie rural ; N° 1- 2, 6168p.

47. Nolasco, R. (2013) Evaluation of the current contamination of heavy metals and certain compounds of sporting interest in the St. Lawrence River in Quebec City. Essay presented at the Centre universitaire de formation en environnement en vue de l'obtention du grade de *maître en environnement (M. Env.)* université de Sherbrooke: p.34.

48. OUALI Naouel 2018 Identification and quantification of a trace metal matrix in the marine

environment

49. Pandaré, D. and Tamoïkine, M.Y., 1994: Etude de l'ichtyo plankton dans les eaux côtières et estuariennes bordées de mangrove en Guinée et au Sénégal. In: Dynamique et usage de la mangrove dans les pays des rivières du sud (du Sénégal à la Sierra Leone). Marie Christine, Cormier Salem, ORSTOM.

50. Paugy et Metayer, C. 2003 La fécondité des poissons téléostéens. Collection de Biologie des Milieux Marins. 5th Ed, Masson, 121 p.

51. Pezennec, O., 1999: L'environnement hydro - climatique de la Guinée. In: La pêche côtière en Guinée: ressources et exploitation. Paris: (ed. Sci.): IRD / CNSHB, p. 16

52. Pichard A., Bison M., Diderich R., Doomaert B., Lacroix G., Lefevre J.P., Leveque, Magaud H., Morin A., Oberon D., Pepin G., Tissot S., 2003-Plumb and its derivatives. Fiche de données toxicologiques et eenvironnementales des substances chimiques.

53. Picot André, 2002. European expert in toxicology. The trio of mercury, lead and cadmium. Les métaux lourds : de grands toxiques.2002.

54. Prankel, S.M.; Nixon R.H.; Philips, C.J.C. (2004). Meta-analysis of feeding trials investigating cadmium accumulation in the livers and kindeys of sheep. Environmental research 94,171,183. Revue de medecine interne ,17:826-835.

55. Ramade F, (2000): Dictionnaire encyclopédique des pollutions. Ediscience International, France, 690p, p428.

56. Rocher, V. (2003). Introduction et stockage des hydrocarbures et des éléments métalliquesau sein du réseau d'assainissement unitaire parisien (Doctoral dissertation, Ecole des Ponts ParisTech).

57. Rodier, J. (1984). L'analyse de l'eau: eaux naturels, eaux résiduaires et Eaux de mer. 7th edition. Paris. 1365 p.

58. Safaa FOUAD , Kaoutar HAJJAMI , Nozha COHEN et Mohamed CHLAIDA-Qualité physicochimique et contamination métallique des eaux de l'Oued Hassar : impacts des eaux usées de la localité de Mediouna (Périurbain de Casablanca, Maroc) *Afrique* SCIENCE *10(1) (2014) 91 - 102*

59. SAHBAOUI Fatiha 2015 Contribution to the study of contamination by some heavy metals in the fish *Dantex engolensis* at the Ghazaouet coast (Wilaya of Tlemcen)

60. Schoeters G., 2006. Cadmium and children: exposure and health effects. Acta Paediatr. 95 (Suppl.),

61. Theopkile R. Camara I N. Diallo S T. (2006). Rapport Nationale Sur l'environnement Marin et Côtier.

62. Thierno Sadialiou BAH Effects of human activities on the dynamics of the Tabounsou-Conakry estuary

63. Türkmen, A., Türkmen, M., Akyurt, I. (2005). Heavy metals in three commercially valuable fish species from Iskenderun Bay, Northern East Mediterranean Sea, Turkey. Food

Chemistry,

64. UNEP/FAO/WHO, 1996-Etat du milieu marin et du littoral de la région méditerranéenne. MAP Technic.Rep. Ser. 101: 1-148.

65. Wassim G. (2017). Pollution and Nuisance. Gabe.

66. WHO, 2004; UNEP, 2005; EC, 2011; Survival of pathogens- Final Reports on Research Projects?

67. WOGNIN S. B. (2008) : Publications et communications, Abidjan, Université de Cocody, UFR des Sciences médicales, CAMES, 335 p.

68. Yassine M., & C. Brunot (2011) The coastal and marine environment. Institut français de l'environnement. Études et Travaux n° 16: 116 p.

69. L'Emmanuel P. Dinnat(2003) - the determination of ocean surface salinity using microwave radiometric measurements in band

Webography

https://www.eaufrance.fr/les The impact of water pollution (2020). The impact of water pollution.

Archives

Conakry Governorate Archives 2014

National Directorate of Fisheries and Aquaculture Conakry 2014

Direction nationale du tourisme et hôtellerieConakry,2014

Direction Nationale de la Station Meteorologique-Conakry, 2014

Direction de l'environnement eaux et forêts-Conakry, 2014

A NNEXES

Table A: FAO/WHO guide values, 1989 and European Community Standards EC 2002 and International Standards for heavy metals in fish muscle, sediments and water.

| Trace metals mg/kg | Sediment | | Fish fabric | | | Water |
	ABRMC	GESAMP	FAO/WHO	Range of International Standards	CE	World Bank
Ar	10	-	-	-	-	-
Cd	0,6	0.11	1	0-2	0.05	0,05
Cu	26	33	30	10-100	-	0,5
Cr	45	-	1	-	-	-
Pb	22	19	0.5	0.5-10	0.5	0,5
Zn	88	95	100	40-100	-	1
Fe	2000	-	-	-	-	-
Mn	400	-	1	-	-	-
Ni	45	-	20	-	-	0,12
Hg	0,2	0.05-0.3	0.5	0.5-5	0.5	-

Table B: Sample collection points and their geographical coordinates.

N°	Sampling sites	Sampling points	Latitude	Longitude
1	Faban	Sea water near the landing stage	9° 51'45.90"N	13° 71' 51,18"W
2	Gbessia port 2	Sea water near the beach (bernarese)	9° 31'18.98"N	13° 41' 57.98"W
3	Gbessia port 1	Seawater near the outlet of the Plant	9° 55'36.22"N	13° 67' 30,82"W

Table C: Comparative representation of the results obtained (physico-chemical) with those of other authors

Algeria					
Locations	T (°C)	S (%)o	pH	O2 (%)_mg/l	References
Skikda Bay	27.09	40.23	8.15	50.06	**Gueddah (2003)**
Port of Algiers	20.55	32.7	6.88	3.2mg/l	**Laama (2009)**
Oran Bay	21.3	34.3	8.21	2.59	**Ouali (2006)**
Mostaganem Bay	18.5	35.8	7.93	2.90	**Remmili and Kerfouf (2013)**

Souk Tlata coast		18.40	-	8.26	-	Hachemaoui (2014)
Annaba Bay		19.46	36.36	7.76	51.86	Ms OUALI Naouel (2018)
Guinea						
Bay of Tabounsou	Conakry Port Authority	24,8	32,4	7,2	5,6	
	Boulbinet	25,02	33,7	6,5	8,8	
	Boussoura	25,08	22,76	8,5	9,7	
	Dabondy	21,61	19	7,8	6	
	Yimbaya Tanerie	28,64	28,55	8,1	5,8	**Dr. Abdoulaye BAH (2019)**
Bay of Landreah	Sangaréah Badè	32,91	22,05	9,6	6,9	
	C.T.Kaloum	37,21	11,71	6,9	5,3	
	Dixinn port3	25,06	23,62	7,43	6,8	
	Kipé	25,07	38,5	5,83	5,5	
	Kaporo Beah	25,04	-	8,84	6,3	
	Sonfonia Port	25,03	-	6,91	7,7	
The Konkouré estuary		31,02	33,04	6,98	6,5	**Onivogui G et *al*, (2013)**
Tabounsou Bay (the average of the results from the three sampling sites)		**26,7**	**30,13**	**7,3**	**7,9**	**This study 2020-2021**

Table D: Comparison of the results of the analysis of metals in water with bibliographical data.

Locations		Zn (mg/l)	Cu (mg/l)	Cd (mg/l)	Pb (mg/l)	References
Tabounsou Bay	Autonomous Port of Conakry	1.98	0.71	7.03	3.07	
	Boulbinet	1.18	0.075	6.5	2.33	
	Boussoura	0.84	0.93	0.08	0.19	
	Dabondy	1.58	0.76	0.69	0.26	
	Yimbaya Tanerie	1.54	0.95	8.06	2.42	**Dr. Abdoulaye BAH (2019)**
Sangaréah Bay	Sangaréah Badè	0.23	0.019	0.25	0.14	
	C.T.Kaloum	2.48	0.58	6.9	3.88	
	Dixinn port3	1.28	1.61	0.37	3.81	
	Kipé	1.62	0.079	8.5	5.17	
	Kaporo Beah	1.35	0.22	0.91	2.87	
	Sonfonia Port	1.99	0.35	8.05	2.39	
The Konkouré estuary	Kakounsou	180	0.80	0.025	1	**Onivogui G et *al*, (2013)**
Tabounsou Bay (the average of the results from the three sampling sites)		**1.47**	**1.07**	**0.027**	**0.12**	**Current study (2020-2021)**

Table E: Comparison of results of metal analysis in fish fillets with bibliographical data.

Locations	Species	Heavy metals				Reference
		Zn	Cu	Cd	Pb	
Lake Manzala (Egypt)	*Mugil cephalus*	29.62	5.75	2.75	1.74	**Bahnasawy et al., 2009**
Tuzla Lagoon (Turkey)	*Mugil cephalus*	39.6	0.53	0.92	0.09	**Dural et al., 2007**
Monolimni Lagoon (Turkey)	*Mugil cephalus*	220	50	3.0	0.6	**Boubonari et al., 2009**
Bejaia Algeria	*Dentex angolensis*	0,239	0,685	0,010	-	**Mansouri Kahina et al, 2016**
Ghar El Melh lagoon (Tunisia)	*Mugil cephalus*	-	0.202	0.09	0.338	**Chouba et al., 2007**
Lake Qarum ... Egypt)	*Mugil cephalus*	2.69	2.73	-	-	**Authman & Abbas, 2007**
Mediterranean Turkey	*Pomadasys jubelini.*	34,58	4,17	0,55	5,57	**Canli et al, 2002**
Iskenderun Bay (Turkey)	*Mugil cephalus*	0.005	0.001	0.002	0.0003	**Turkmen et al., 2006**
Karatas coast (Turkey)	*Mugil cephalus*	24.2	9.8	6.3	1.5	**Çogun et al., 2006**
Tabounsou Bay (Max-Min size)	*Pomadasys jubelini*	(70-61,3)	(22 -19)	(0,4- 0,34)	(0,04 - 0,016)	**Current study (2020-2021)**
	Mugil cephalus	(30-26)	(32,5-31)	(0,6-0,49)	(0,07 -0,055)	
	Dentex angolensis	(24,5- 22,8)	(29,6- 28)	(0,21-19)	(0,067 -0,03)	

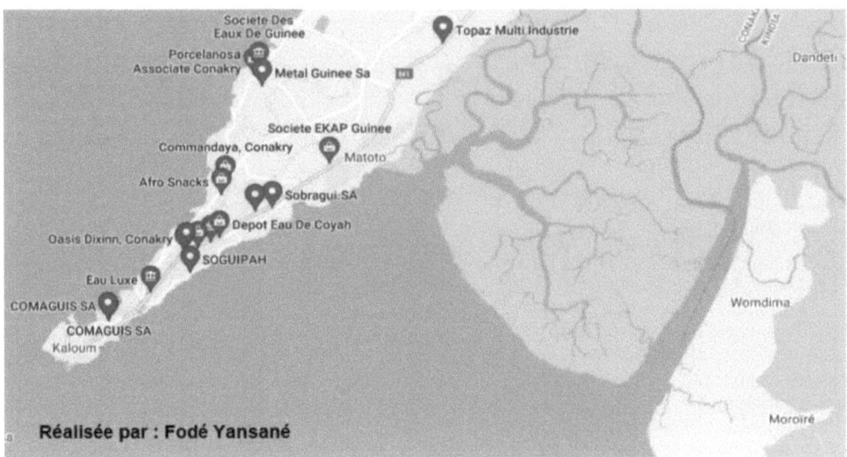

Fig 1: Map of some of Conakry's factories

Fig 2: Unplanned buildings in Faban (south of Conakry)

Fig 3: Discharge of liquid effluent from the Kaloum thermal power plant into the sea

Fig 4: CTK main collector leading effluent into the lake and out to sea

Fig 5: Channel for discharging mixed waste from the Tannerie to the sea

APPENDICES III: SURVEY FORMS

A.TO THE AUTHORITIES

---------- Date/2020Port number   : ------------------

Respondent : ____

How many people are in charge : ---

Nationality: Guinean   □; Leonese   □; Ghanaian □; other O

QUESTIONS

Are there any human activities along the Tabounsou estuary?

Yes □No   □I   have no idea □

What are the Most Popular?

....

....

....

Is there a link between your department and the various activities on the Tabounsou estuary?

Yes □No □

I have no idea □

Do you carry out surveys along the estuary?

  Yes   □No □I   have no idea □

Do you have any statistical data on these riverside activities?

Yes □Nol   I have no idea □

Do these activities have an impact on the estuary?

Yes   □No   □I   have no idea □

What effects did you observe?

....

....

....

Which species are landed most often?

....

....

....

B.TO FISHERS

       ---------- Date/2020   Port ------------ number ----------------------:

Respondent : --

Main activity: ----------------------------- Secondary activity :   -------------------------

Nationality: Guinean   □; Leonese   □; Ghanaian □; other □

Questions

Do you use fishing gear? Yes No

If so, what fishing gear do you use?

    ✓ Set gillnet (FMC)   yes □   no □
    ✓ Drift gillnet (FMD)   yes □   no □
    ✓ Encircling gillnet (FME)   yes □   no □
    ✓ Revolving net   (FT)   yes □   no □
    ✓ Line   (LI)   yes □   no □
    ✓ Longline   (PA)   yes □   no □

Does this machine meet the standards? Yes O No □

If so, what mesh size do you use   and what type of wire do you use?

Do you use the boats?

Yes...   No ...

If so, which boats do you use?

Is this estuary rich in fish stocks?

Yes... No ...   I have no idea □

Which species are caught the most?

What is the quantity of species fished per day in kilograms?

Fish : --------- Shrimp : ------- Crab : -----------

Are there any species threatened with extinction Yes ... no □ I have no idea ...

If so, which ones?

....

....

....

What do you think is behind this threat?

....

....

....

Has there been a change in your catches compared with previous years?
Yes ...   No ... I have no idea ...

If so, what do you think is responsible for this variation?

....

....

....

As fishermen in the Tabounsou estuary, what are your main difficulties?

....

....

....

....

....

C.TO FARMERS/RIDERS

_______ Date/2020   Card no.: ----- Port : --------------------------------
Respondent :

Main activity: ------------------------------ Secondary activity :   -------------------------

Nationality: Guinean   □; Leonese   □; Ghanaian □; other O

Questions

How big is your field?
....

For how long can the bank be cultivated?
....

Do you sometimes increase the surface area of your field?

Yes   ....Non ... I have no idea ...

Do you use chemicals in agriculture? Yes ... no ...

If yes

Which : ....

How much do you use? ....

Is bank farming profitable at all times?

Yes ...No....I   have no idea I I

What do you think could be the causes of any variations?

....

....

....

What do you see as the future of riverside farming?

....

....

....

As a farmer/riparian on the riverbank, what are your main difficulties?

....

....

....

Printed by Books on Demand GmbH, Norderstedt / Germany